University of Bath • Science 16–19

Series Director: Prof. J. J. Thompson, CBE

Microorganisms and Biotechnology

SECOND EDITION

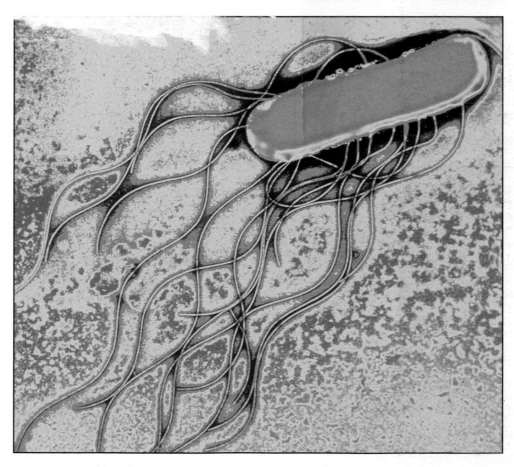

JANE TAYLOR

First published in 1992 by:
Thomas Nelson and Sons Ltd

Second Edition published in 2001 by:
Nelson Thornes Ltd
Delta Place
27 Bath Road
CHELTENHAM
GL53 7TH
United Kingdom

01 02 03 04 05 / 10 9 8 7 6 5 4 3 2 1

A catalogue record for this book is available from the British Library

ISBN 0 17 448255 8

New Illustrations by Oxford Designers and Illustrators
Page make-up by Jordan Publishing Design

Printed and bound in Italy by Canale

Acknowledgements
The author and publisher are grateful to the following for permission to reproduce photographs:
Cover photo: Science Photo Library
Agricultural Genetics Company Limited 105; Heather Angel/Biofotos 24 (top); Biophoto Associates 8 (left & right), 24, 161 (bottom); Biotechnology and Biological Sciences Research Council 100; Anthony Blake Photo Library 67; Bruce Coleman 33, 165; P Dunnill 109; Farmers' Weekly Picture Library 160; Geoscience Features Picture Library 161 (top); Holt Studios International (Nigel Cattlin) 112 (both); John Innes Centre 167 (top); MAFF 166; Mark Moodie 81; National Dairy Council 66; Ann Ronan 4; Science Photo Library 1 (© Alfred Pasieka/SPL), 7 (left, centre & right), 13 (© CDC/SPL), 14, 21 (© Eye of Science/SPL), 25, 27, 38 (© Hank Morgan/SPL), 51 (© Hank Morgan/SPL), 89, 93, 107 (© Dr Jeremy Burgess/SPL), 116 (© Makoto Iwafuji/ Eurelios/SPL), 119 (© Hans-Ulrich Osterwalder/SPL), 124, 129,147, 154 (© London School of Hygiene/SPL), 161 (middle), 164 (© Astrid & Hanns-Frieder Michler/SPL), 168 (© Peter Menzel/SPL), 170 (© Rosenfeld Images Ltd/SPL); South American Pictures 87; Stilton Cheese Makers Association 58; C James Webb 39, 47, 54, 99, 145, 151; Wellcome Trust Photo Library 14 (both), 107 (top); Wildlife Matters 79; World Health Organisation 131, 135; Zefa Picture Library 57
Picture research by Sue Sharp

AQA (AEB/NEAB) examination questions are reported by permission of the Assessment and Qualifications Alliance; OCR questions are reproduced by permission of OCR.

Other titles in the Project

Physics (2nd Edition)
Robert Hutchings

Astrophysics
Neil Ingram

Electronics
Jackie Adams and Robert Hutchings

Energy
David Sang and Robert Hutchings

Nuclear and Particle Physics
Brian Cooke and David Sang

Telecommunications
John Allen

Biology
Martin Rowland

Applied Ecology
Geoff Hayward

Applied Genetics
Geoff Hayward

Biochemistry and Molecular Biology
Moira Sheehan

Environmental Science (2nd Edition)
Kevin Byrne

Medical Physics
Martin Hollins

Chemistry
Ken Gadd and Steve Gurr

Contents

Bath Advanced Science Series: Introduction

It is now some ten years since the **University of Bath 16–19 Project** was first published. The success of the material produced by the team of talented authors has been demonstrated consistently throughout that period. At the same time there has been a great deal of change in the provision of education, not only in the UK but throughout the world. With the ongoing drive towards the broadening of post-16 education – making it *more* accessible and *more* useful to *more* people – there is greater emphasis not only on expanding knowledge, but also on learning and understanding how this knowledge can be applied, and developing the skills required to do so.

The **Bath Advanced Science** series represents an extensive review, re-organisation, and update of the material produced in the original project. The author team has come together to draw on its collective expertise and experience to produce resources which reflect the demands of post-16 education today.

The series has been written to address the Advanced Subsidiary and Advanced GCE Science specifications introduced by the unitary awarding bodies from September 2000 following the Dearing *Review of Qualifications for 16–19 Year Olds* and the subsequent consultation *Qualifying for Success*.

- The subject texts provide comprehensive coverage of the content required for the awarding body specifications in those subjects.
- The topic books each provide detailed coverage which will support both option topics for AS and A GCE and other advanced level courses. They are also appropriate for students embarking on Foundation courses in Higher Education, or preparing for degrees in which these topics are studied.

The **Bath Advanced Science** series contains up-to-date material presented to encourage effective learning. The texts continue to be highly interactive – ensuring a degree of independent self-study and actively providing opportunities for students to acquire knowledge and understanding, develop skills and concepts, and to appreciate the applications and implications of science.

We have designed the **Bath Advanced Science** series specifically to help all students to develop the realisation that science *is* for everyone and to gain an appreciation and understanding of the importance and relevance of science to everyday life.

Prof. J J Thompson, CBE
Department of Education
University of Bath
May 2000

How to use this book

The aim of this book is to describe the nature, importance and applications of microorganisms and cell technology. This is an area of biology which has developed very fast in recent years and has an increasing impact on everyday life. Its growing importance is reflected in the greater emphasis given to these topics in specifications for Curriculum 2000. The material covered fits the requirements of the AS and A2 specifications and complements your core biology text in extending your knowledge and understanding of a rapidly developing subject.

Microorganisms and Biotechnology is written for students who are following AS and A2 courses in Biology, or Human or Social Biology. It will be helpful to those following GNVQ Health Studies or Science, and Home Economics A-level.

There are three themes: the first deals with the nature and growth of microorganisms; the second with environmental and industrial microbiology, and the final theme covers microorganisms and disease in plants and animals, together with plant biotechnology.

Although it may be easier to start at the beginning of the book and work your way through, you don't have to do this. You can use just a couple of chapters or a single theme without having to study all the other parts, but if you do so, you should read some sections of Chapters 2 and 3 first. These sections are listed on the pages introducing each theme, under the heading prerequisites. The chapters will cover more topics than in your specification and more than you are likely to have to learn, but even if you do not need to study a particular section in depth you will find that reading through it helps to broaden your understanding. You should consult your specification to see which topics you need to study in depth and you will find the index and contents list help to locate a particular item.

The material is arranged to assist you in organising your studies. On the page opposite you will find information about the different features which will help you get the best out of the book.

In examinations you are expected to be able to select and use correct scientific terminology, and may be penalised if you do not do so; the material is written to help you understand and use technical terms. When an important word or technical term is used for the first time it is printed in heavy type, **like this**. It may be defined, or its meaning explained in the surrounding two or three sentences. It is not explained again if it is used later in the book. If you are just using one or two chapters of the text and come across a technical term you don't understand you can use the index. There is also a glossary of some biological terms which may be new to you.

I would like to thank the many students and teachers who have used the first edition of this book and sent me their invaluable comments and suggestions. In particular, I would like to thank my own A-level students who have patiently endured being guinea pigs for all kinds of exercises and ideas. When you use this text I hope that you not only learn a lot but also get some enjoyment out of reading about the world of microorganisms and how we exploit them.

Learning objectives

Each chapter starts with a list of learning objectives which outline what you should know and understand after studying each chapter. Those starting with the words 'describe' or 'explain' indicate that you must remember, be able to use, and give reasons for the information indicated. Learning objectives can help you revise, especially if used in conjunction with the summaries at the end of the chapter. Please check your specification or see the website on www.bathscience.co.uk to establish which topics you have to study.

Questions

There are questions at the end of sections of work. Some test recall and need only a word or phrase in answer; others need a longer explanation, or require you to convert information from one format to another. These help you memorise the facts that you need. Others ask you to use knowledge to interpret data, and some are open-ended with many possible answers. The questions start simply and become more demanding. Some questions enable you to provide evidence for Key Skills. You can do the questions as you go along while the information is fresh in your memory, or save them to do at the end of the section; but remember the longer you leave them the harder they are to answer. Answers, or answer outlines, are provided on the website for your guidance where appropriate.

Case studies and Spotlight boxes

These are extension topics related to work in the main text. You may not have to learn these but they have been included to help your understanding of microbiology, its applications and areas of increasing importance not yet in specifications.

Summaries

Each chapter ends with a brief summary of content. These summaries, together with the learning objectives, should give you a clear overview of the subject and allow you to check your own progress.

Investigations

Though some practical techniques are illustrated there are no specific practical exercises. However, many of you will have to carry out an independent investigation or coursework, so there are some ideas at the end of the book. Some of these do not require bench practical work. Activities using the Internet and opportunities for Key Skills evidence are also included.

WHAT ARE MICROORGANISMS?

Microorganisms are a very diverse group of organisms linked only by their small size. However, they are important in every area of life. The invisible world of microorganisms has been explored since the invention of the microscope but real understanding has only developed in the last 150 years. New information about their metabolism is arriving every day. They have made us rethink many ideas about life and how life has evolved; even the definition of life as the boundary between a living organism and genes or proteins becomes blurred. As more and more biotechnological processes become mainstream commercial activities, we need to have a better understanding of microorganisms and their growth. Chapters 1 and 2 introduce you to the groups of microorganisms, while Chapter 3 tells you how we grow them. ■

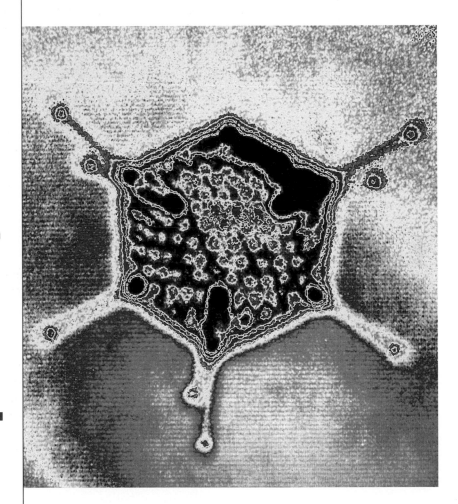

PREREQUISITES

An understanding of GCSE chemistry; it is advantageous to have studied A-level cell biology and respiration.

1 INTRODUCING MICROORGANISMS AND BIOTECHNOLOGY

LEARNING OBJECTIVES

After studying this chapter you should be able to:

① appreciate the main steps in the development of knowledge of microorganisms

② appreciate the size and use the correct units to measure microorganisms

③ describe the structure of a prokaryotic cell

④ describe the differences in structure between prokaryotic and eukaryotic cells.

WHY DO WE STUDY MICROORGANISMS AND BIOTECHNOLOGY?

1.1

icroorganisms have an enormous impact on our lives. They make changes to our environment; they cause animal, plant and human diseases; and they synthesise important commercial and medical products. Our understanding of how microorganisms work and of their abilities has come from scientific investigation of their activities. Information gained by growing microorganisms has helped us grow isolated animal and plant cells in similar ways.

Biotechnology uses microorganisms, their products, and cells in industrial processes to make commercial products and provide services. Everyone's life is touched by microbial technology in some way. This morning you may have dressed in clothes washed in washing powder containing microbial enzymes and eaten breakfast from plates washed in a liquid also containing them. The water you used to wash with was cleaned up by the action of microbes. Your toast was made with yeast – another microbe. Later in the day you may eat chicken nuggets with a coating containing genetically modified soya, or a meat-free meal containing fungal protein products, or cheese made with microbial rennet, and a low fat dessert thickened by microbial gum, possibly washed down with beer or wine – made by microbes. The pretty pansies growing in the garden were micropropagated, and so were some of the flowers in the Mother's Day bouquet. You might even have slept on sheets made from genetically modified cotton. You are protected from a range of diseases by vaccines made using microbes, and take antibiotics made by microbes; you might have had a blood test using enzyme technology, or a pregnancy test using monoclonal antibodies. The purity of products you use and your food and water may be monitored by biotechnological devices. In some parts of the world you might drive a car using fuel made by microbial fermentation of agricultural waste, or depend on lighting and heating fuelled by gas produced in microbial fermenters. New genetic techniques have led to further advances in medicine, agriculture, food technology and commerce. As our knowledge has increased, biotechnology has widened to include many more processes involving cell products such as enzymes, cells and tissues, ranging from genetic fingerprinting to techniques for gene therapy. This book cannot cover the whole range of activities but focuses on and explains more about each of the processes mentioned above.

1.2 THE HISTORY OF BIOTECHNOLOGY

FIG 1.1 Louis Pasteur is regarded as the father of medical microbiology.

We know that microorganisms were used thousands of years ago in the Middle East for brewing and baking, but they are too small to be seen and were not recognised as living things at the time. Very small organisms were first observed when the microscope was developed in the 17th century.

Great advances in our understanding of microorganisms were made in the 19th century. People learned that fungi could cause plant diseases and preventative treatments were developed. We also gained a better understanding of the role of microorganisms in disease. Most of the credit goes to two men and the scientists they gathered around them. Louis Pasteur, shown in Figure 1.1, was French; Robert Koch, slightly younger, was German. Their governments had been at war when they were both young men and they met only once. Joseph Lister, the British surgeon who introduced the use of antiseptics, arranged the meeting but it was reported as being a dismal failure, and they worked in great rivalry and with some animosity for the whole of their lives. However, between them, they can be credited with

- the identification of the organisms causing many diseases;
- the understanding of how microorganisms cause illness;
- developing ways to grow, examine and investigate microorganisms still in use today;
- the systematic development of vaccines and anti-toxins;
- the development of hygienic practices to prevent the spread of infectious diseases.

By the start of the 20th century the role of microorganisms in the decay and nitrogen cycles became clearer and viruses were discovered. Major steps in our understanding of microorganisms and biotechnology in the last century can be seen in Table 1.1. The 1970s and 1980s led to a huge growth in the exploitation of cells, microorganisms and their enzymes to carry out a variety of industrial production processes more cheaply and more effectively than ever before. By the 1980s genetic engineers could transfer genes between species to improve crop productivity and give microorganisms and animals new biochemical abilities. Genetically modified crops were grown commercially by the end of the century.

TABLE 1.1 A time line of landmarks in microbiology

17thC	**Anton van Leeuwenhoek**, 1632–1723, develops a microscope and discovers microorganisms. 1683 Mouth bacteria drawn.
1718	**Lady Mary Montague** introduces inoculation against smallpox from Middle East.
1796	**Edward Jenner** experiments with a cowpox vaccine to prevent smallpox.
1802	Forsyth, the King's gardener, uses lime sulphur to treat mildew on fruit trees.
1846	Rev. Berkeley states that a fungus caused potato blight in the Irish famine.
1856	**Louis Pasteur**, 1822–1895, establishes that yeast is needed for fermentation.
1857	Pasteur discovers that wine souring is caused by microorganisms found in vinegar barrels. Develops pasteurisation. **Anton de Bary** shows potato blight to be caused by a fungus.
1864	Pasteur establishes that decay organisms are found as small organised 'corpuscles' or 'germs' in the air.
1865	Pasteur investigates silkworm disease and establishes that diseases can be transmitted from one animal to another. **Joseph Lister** reads Pasteur's work and uses disinfectants in surgery.
1866	Pasteur develops a germ theory of disease.
1868	Davaine uses heat treatment to cure a plant of bacterial infection.
1873–76	**Robert Koch** investigates anthrax, and develops techniques to view, grow, stain and photograph organisms, aided by Gram, Cohn & Weigart.
1879	Pasteur grows weakened organisms that cannot cause disease but protect against severe strains.
1880	Koch establishes a protocol to show that an organism causes a disease.
1882	Pasteur uses Koch's work to produce a vaccine against anthrax. Koch identifies the TB organism and hence the first human microbial disease. **Ilya Metchnikoff** observes phagocytes in action. Develops a cell theory to explain the action of vaccines. Millardet develops Bordeaux mixture to treat powdery mildew on imported vines.
1883	Hansen isolates a yeast, *Saccharomyces carlsbergensis*, used in brewing.
1885–95	Koch, **Petri**, Löffler, Yersin & **Erlich** identify many human disease-causing organisms. Behring develops diphtheria anti-toxins.
1886	J. C. Arthur demonstrates that pear fire blight is a bacterial disease.
1890s	Roux develops diphtheria vaccine.
1892	**Iwanowski** identifies tobacco mosaic disease as a virus disease.
1898	Löffler & Frosch identify the first animal viral disease, foot-and-mouth disease.
1900	**Walter Reed** identifies yellow fever as the first human viral disease.
1906	Paul Erlich works on atoxyl compounds and finds Salvarsan – the first chemotherapeutic agent.
1908	**Calmette** & **Guerin** develop a vaccine against TB, BCG used in 1921.
1913	First enzyme-based washing powder produced.
1915	**Twort** describes bacteriophages: viruses that attack bacteria
1928	**Alexander Fleming** observes that bacteria are inhibited around a colony of *Penicillium* mould. **Fred Griffith** transforms bacteria – they gain a new capacity to make something.
1930s	**Gerhard Domagk** investigates a dye, Prontosil. Later the Trefouels derive a sulphanilamide drug. Isolated plant tissues cultured. Organo-sulphur fungicides developed.
1935	**Stanley** crystallises tobacco mosaic virus.
1938	**Florey** & **Chain** isolate the anti-microbial agent penicillin.
1939	**Gauteret** grows carrot callus cultures.
1946	**Lederberg** & **Tatum** discover how bacteria transfer genes.
1950s	Earle & Enders grow monkey, mouse and chick cells in cell cultures. Gey develops the HeLa human cell line.

TABLE 1.1 *continued*

1959	**Reinart** regenerates plantlets from carrot callus culture.
1961	**Jacob** & **Monod** establish the basis of genetic regulation. Systemic fungicides developed.
1963	Biological washing powder launched in Europe.
1966	DNA code completely deciphered.
1970	Arber, Nathans & Smith work on restriction enzymes which snip DNA. **Temin** & **Baltimore** discover reverse transcriptase which makes DNA from RNA.
1960–70s	Plant cell protoplasts cultured. Viroids identified.
1965	Harris & Watkins fuse mouse and human cells together.
1968	**Edgar** *et al.* determine how bacteriophage T$_4$ infects *E.coli*.
1970	**Khorana** *et al.* synthesise an artificial gene.
1972–73	Boyer & Cohen splice DNA into a plasmid and pass it into *E. coli*.
1975	**Kohler** & **Milstein** fuse cells together to produce monoclonal antibody. Voluntary guidelines to control gene transfer experiments agreed.
1977	Itakura and colleagues engineer a bacterium to make somatostatin, followed by human insulin and growth hormone.
1980	US courts allow patent on modified bacterium.
1982	First genetically modified mice produced.
1983	**Kary Mullis** develops PCR, which enables rapid gene copying.
1984	**Alec Jeffreys** and team discover the technique of genetic fingerprinting.
1983–87	Field trials of genetically modified plants. Human insulin produced by yeast.
1988	First genetically engineered oncomouse patented.
1994	BST milk on sale.
1995	Flavr Savor genetically modified tomatoes on sale. The first complete DNA sequence for an organism, *Haemophilus influenzae*, published.

1.3 WHAT ARE MICROORGANISMS?

'Microorganisms' is an arbitrary grouping, defined as organisms that are too small to be seen unaided. This is not always true. For example, we can see mushrooms, which are fungal fruiting bodies, and the thread-like structures of fungi growing through decaying leaves or food. Microorganisms include bacteria, fungi, viruses, protozoa and green algae. In this book the emphasis is on those microorganisms that are important for industrial, commercial or health reasons.

Microorganisms colonise all ecological niches, including hostile places where few other living things can survive such as the Arctic, inside engine fuel lines, coal mines and hot springs. They play key roles in photosynthetic productivity, in decay and as parasites of all groups of living things. Most microorganisms are **unicellular,** that is, they are a single cell that carries out all life functions: feeding, respiring, reproducing, excreting metabolites and so on. Most microorganisms cannot move about independently but some have structures that give them mobility. There are **multicellular** microorganisms, for example the fungi, which may make specialised reproductive cells.

1.4 SEEING MICROORGANISMS

We can see objects as small as 0.1 mm across without a microscope but microorganisms are usually even smaller. An optical microscope with magnifications of up to ×2000 allows us to see things as small as 0.2 μm. It reveals the external features of bacteria and protozoa. We learn more about structure by using particular stains that are taken up by particular cell components. With stains we can distinguish cell walls, an outer cell membrane, structures such as cilia and flagella, and large cell organelles such as nuclei, chloroplasts and stained mitochondria but no details of their internal structure.

The internal structure, or ultrastructure, of cells is seen using an electron microscope. An electron microscope uses a beam of electrons with a shorter wavelength than light and allows us to see objects a hundred times smaller than an optical microscope. It will resolve structures as small as 0.3 nm. Different techniques show different structural details. For example, freezing then fracturing a specimen may split it along membranes, revealing the inner structural planes. A scanning electron microscope forms an image from electrons scattered by the surface of a specimen and reveals something of the three-dimensional structure of a specimen. Compare the images from each technique in Figure 1.2.

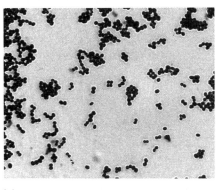 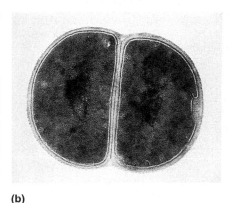 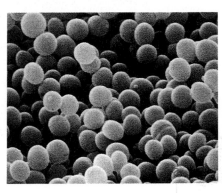

(a)   **(b)**   **(c)**

FIG 1.2 The same organism viewed by three different techniques: **(a)** is a smear of *Staphylococcus pyogenes aureus* stained and photographed using a light microscope, originally ×500. No internal detail can be seen. An electron microscope was used to take **(b)**, a newly divided *S. pyogenes aureus* cell with an original magnification of ×30 000. The wall and some internal detail can be seen. A scanning electron microscope was used for **(c)** at a similar magnification to **(b)**. The shape and surface detail of the bacteria can be seen.

Units

Microorganisms are usually measured in micrometres (μm). One micrometre is one thousandth of a millimetre, or one millionth of a metre (10^{-6} m). Animal and plant cells, protozoa and algae are between 0.5 and 20 μm in diameter and can be seen with an optical microscope. Even large bacteria are less than 4 μm long and the viruses are at least ten times smaller than this. For these very small microorganisms the nanometre (**nm**) is used. This is one millionth of a millimetre (10^{-9} m).

1.5 NAMING MICROORGANISMS

Some microorganisms have common names from their links with human life, for example yeast or measles virus. Most don't and are known by a two-part scientific name which reflects links with other microorganisms. Microorganisms are grouped using structural features and components held in common. We still have gaps in our knowledge so some organisms are grouped with those with which they have most similarities. As we learn more about an organism's metabolism, and particularly its genes, we can assign it more accurately and may have to rename it.

The first part of the name is its genus (written with an initial capital letter) and it is shared with close relatives. The second part of the name is unique to a species and is based on features such as its metabolism, source, its discoverer or any disease it causes. Some species have a third part to the name which refers to a particular strain, e.g. *Escherichia coli* O157. Closely related species may be referred to by just the genus name when most share a feature, for example '*Lactobacillus* can tolerate acid conditions'. Viruses are usually named after the disease they cause and the organisms affected. For brevity microorganisms are referred to by a shortened version of their name, for example *E. coli*.

QUESTIONS

1.1 List as many ways as you can in which microorganisms affect you in your daily life. Revise the list after you have read more of the book.

1.2 List the following units in decreasing order of size, with the appropriate symbol: nanometre, metre, centimetre, micrometre, millimetre.

1.3 Find out about the conflicts about the nature of microorganisms by reading about the work of Spallanzani and Louis Pasteur on spontaneous generation. Make a short presentation about the debate on spontaneous generation.

1.6 PROKARYOTES AND EUKARYOTES

Cells, including microorganisms, are divided into two major types based on their internal structure. The key feature is the presence or absence of internal membranes. **Eukaryotes** contain structures made of membranes, **prokaryotes** do not. The modern descendants of the earliest living organisms do not have membranous structures such as chloroplasts or mitochondria. These prokaryotes include bacteria and blue-green bacteria (cyanobacteria). The lack of membrane structures can be seen in Figure 1.3. The oldest known fossils at about 3 billion years are stromatolites, which appear to be composed of cyanobacteria and other sorts of microorganisms living in colonies.

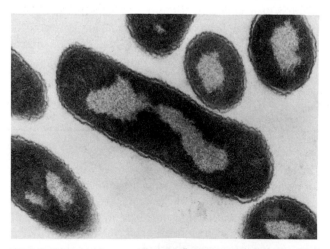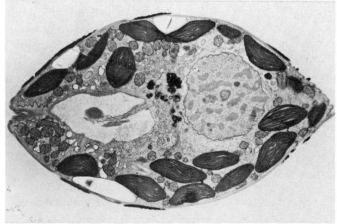

FIG 1.3 Eukaryotes have a different cell structure to prokaryotes. *Escherichia coli* (**a**) is prokaryotic – there is a clear cell wall and a paler area of nucleoplasm. In contrast, *Euglena gracilis* (**b**) is eukaryotic with a clearly defined nucleus, chloroplasts, mitochondria and internal membranes.

Organisms with eukaryotic cells include protozoa, fungi, algae and all more complex organisms. The cells contain characteristic organelles that compartmentalise cell activities and allow more metabolic activities to take place simultaneously without interfering with each other. Eukaryotic cells are between 1 and 100 μm in diameter, which is ten times larger than a typical prokaryotic cell.

TABLE 1.2 Comparison of prokaryotic and eukaryotic features

EUKARYOTIC CELLS	PROKARYOTIC CELLS
Surface membrane	
A phospholipid bilayer with proteins and glycoproteins in a fluid mosaic. Controls the entry and exit of molecules which enter or leave by diffusion, or active transport. Large molecules can only pass by endocytosis and exocytosis. Projections such as microvilli enlarge the surface area.	Also a bilayer. It may be folded inwards, enlarging the surface available for materials to exchange. Functions such as respiration and photosynthesis are located on it. Passage of materials is controlled by the membrane but large molecules such as fragments of DNA and enzymes pass through. Endocytosis is not observed.
Cell wall	
Made of cellulose when present, except in fungi which have cell walls made of a number of substances including chitin.	Most have a thick cell wall; murein is a major component. A barrier and strengthener resisting water entry by osmosis. Not always easy to tell where the cell wall ends and the membrane begins.
Internal membrane structures	
A network of membranes called the endoplasmic reticulum subdivides cytoplasm. It is continuous with the surface membrane and with membranes around the nucleus. Materials pass through these by diffusion and active transport. Regions are studded with ribosomes.	No internal membranes. 70S ribosomes are smaller and scattered, sometimes clustered, in the cytoplasm.
Membrane bounded bodies	
Several types, each specialised for particular functions. Mitochondria are specialised for energy production. Chloroplasts house photosynthesis. Chlorophyll and other pigments are located in thylakoid membranes and the enzymes for carbon dioxide fixation and ATP production are between the layers. Golgi apparatus in secreting cells package materials for transport elsewhere or for release from the cell.	No membrane bounded bodies. Some photosynthetic bacteria and blue-green bacteria may have simple sacs close to surface membranes housing photosynthetic pigments.
Nucleus	
Surrounded by a double membrane with pores. It houses chromosomes, which carry the cell genome. Chromosomes are long molecules of DNA packaged around histone proteins. The nucleus is the site of RNA synthesis and editing. Mitochondria and chloroplasts also carry DNA. The genome is duplicated before mitosis. Genetic exchange occurs during sexual reproduction.	No separate nucleus. The genome is a loop of double-stranded DNA but no histones. It carries the genes needed for growth and development. Many bacteria have separate loops of DNA called **plasmids** carrying non-essential genes. Duplication of the genome and cell division may not be closely linked. Genetic exchange is not linked with reproduction but involves one organism taking up genetic material donated by another.
Osmotic control	
Protozoa regulate the water potential of the internal cell environment by the formation of intracellular vacuoles. These contain surplus water that is expelled to the external environment. Multicellular organisms use different methods.	Osmotic vacuoles are not made by prokaryotes. The cell wall offers resistance to the influx of water.
Reproduction	
Some can use asexual methods without genetic exchange. Sexual reproduction involves the exchange of genetic material. The most common pattern is the production of specialised cells containing half the genome, **gametes**, by meiosis. The genome is completed when two gametes fuse. Many eukaryotic microorganisms are haploid for most of their life cycle. In these, fusion of the gametes produced by mitosis results in a diploid zygote. Then meiosis occurs, making haploid cells again.	Reproduction is usually asexual, taking place by a process of binary fission or budding.

TABLE 1.2 *continued*

EUKARYOTIC CELLS	PROKARYOTIC CELLS
Movement	
Many eukaryotic microorganisms can move: some move by the internal movement of cytoplasm, others have flagella or cilia to propel them through moist environments. Cytoplasmic streaming has been observed in some microorganisms.	Many have no independent movement, though some organisms have a gliding motion. Others have flagella, enabling them to swim through films of moisture.
Flagella	
Some cells have flagella rooted in the cytoplasm projecting beyond the cell membrane and wall. Flagella have a pair of protein fibres extending down the length surrounded by nine pairs of fibres enclosed by membrane. The fibres contract to produce motion.	Flagella have two or three protein molecules entwined to produce a fibre rooted under the cell wall. Acts like a helical rotor to propel the organism through moisture films.

There are other important differences, shown in Table 1.2, between the two groups. Viruses are not included in either group because they do not have a cellular structure. Some viruses have an outer coating of cell membrane but it is acquired from a host as the virus leaves the host cell.

QUESTION

1.4 If you used a light microscope to examine a eukaryotic cell and compared it with a prokaryotic cell, what features would you be able to **see** that would distinguish the two types of cell?

SUMMARY

Most microorganisms are visible only by using an optical or electron microscope. Bacteria can be seen with an optical microscope. The electron microscope reveals details of cell structure.

Microorganisms are measured in micrometres (μm).

There is great variation in size; viruses are as small as a few nanometres whereas the largest bacteria and protozoa are a thousand times larger and measured in micrometres.

There are two types of cells: prokaryotes without internal membrane systems and eukaryotes with complex internal structures. Bacteria and blue-green bacteria are prokaryotic. Prokaryotic organisms lack internal membranes subdividing cell activities and organelles made from membranes.

Fungi, protozoa and green algae are eukaryotic with sub-cellular organelles and a more complex metabolism.

Viruses do not fit into either category, as they do not have a cellular structure.

2. THE MAIN GROUPS OF MICROORGANISMS

LEARNING OBJECTIVES

After studying this chapter you should be able to:

① describe the main features of each group of microorganisms

② explain the main differences between the groups of microorganisms

③ describe how particular examples of bacteria, fungi and viruses obtain their energy and reproduce

④ outline reasons why these groups are important to human well-being.

This chapter is mainly concerned with bacteria, fungi and viruses, but there are other important microbial groups so protozoa and unicellular green alga are also included. Figure 2.1 illustrates the evolutionary relationships of the microorganisms in this book; some links are based on common structures but new links and relationships have been made as we have learnt more about the genes carried by related microorganisms.

2.1 PROKARYOTA

The kingdom **Prokaryota** includes organisms with prokaryote cells, mainly bacteria and cyanobacteria. Bacteria are found in every habitat, growing in a wide range of environmental conditions. Many are saprophytic, that is they degrade organic material in soil and water; others are parasites of animals and plants. But there are also photosynthetic bacteria, and bacteria that gain their energy from inorganic chemical oxidations. In contrast, cyanobacteria are all photosynthesisers. The photosynthesisers in both groups are major primary producers in a number of ecosystems. In favourable conditions both groups multiply quickly and colonise any available substrate.

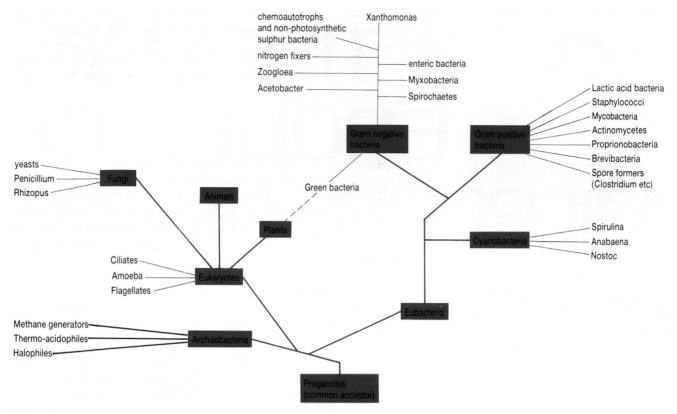

FIG 2.1 The relationships between microorganisms and other groups of living things.

Bacteria are single-celled and range in size from 0.5 μm to 5 μm long. The smallest – the mycoplasmas, included in Table 2.3 on page 29, are only 150–200 nm long. The Actinomycetes are exceptional because they form filaments like fungi. Bacteria were first grouped according to their shape and movement, then subdivided by other features such as growth pattern or the stains they take up. Figure 2.2 illustrates these original groups: the rod-shaped **bacilli**, for example *Bacillus pasteurii* and *Escherichia coli*; the round **cocci**, for example *Staphylococcus aureus* and *Streptococcus lactis*; and the **spiral** bacteria such as *Spirillum* and *Spirochaeta*. These three groups are still useful but have been replaced by two main groupings, each subdivided by metabolic, structural, ribosomal and genetic features.

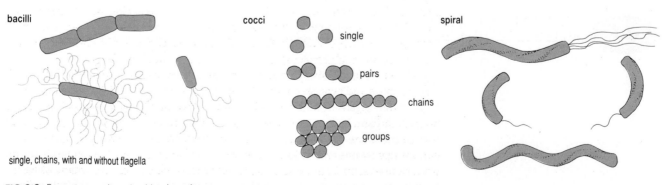

bacilli

cocci
single
pairs
chains
groups

spiral

single, chains, with and without flagella

FIG 2.2 Bacteria are described by their shape.

The **Archaebacteria** are thought to be very old, possibly the earliest life forms, and include the methane-generating bacteria, the halophiles or salt-tolerant bacteria, and the thermoacidophiles which are adapted for life in hot acid places. The **Eubacteria** developed later and include photosynthetic bacteria, sulphur bacteria, slime bacteria, pseudomonads, the enteric bacteria, all Gram-positive bacteria (see below) and the actinomycetes.

Structure

Most bacterial species have a typical shape and size. All except mycoplasmas have an outer **cell wall**, which is different to a typical plant cell wall. The wall gives the cell its shape and maintains its structure. Some species, for example *Streptococcus pneumoniae*, have an outer **capsule** coat. Structures such as **flagella**, seen in Figure 2.3, project through the outer layers. The interior of the cell is filled with cytoplasm but has no organelles such as nuclei or mitochondria. You can see the major features of a typical bacterium in Figure 2.4.

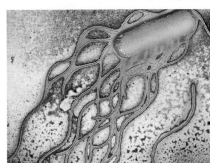

FIG 2.3 Some bacteria can swim using flagella. The salmonella bacteria shown above have flagella projecting from the whole surface.

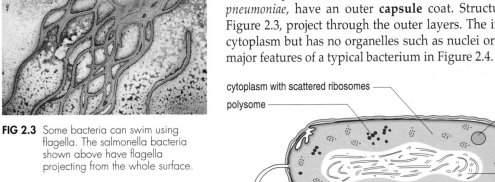

cytoplasm with scattered ribosomes — polysome — storage granule — cell wall — cell membrane — inward fold of membrane — nucleoplasm — carboxysome — mesosome — 1 µm — flagellum

FIG 2.4 The structure of a typical rod-shaped bacterium. Note: only a few species have flagella.

The outer layers

Some bacterial species have a **capsule** acting as a barrier between the cell and its environment. The capsule is usually made of large 'gummy' polysaccharide molecules secreted by the cell membrane and extruded through the cell wall to form a layer around the bacterium. Colonies of capsule-making bacteria are smooth and glistening when grown in a petri-dish. A capsule gives a bacterium an advantage, but it can be washed off without affecting survival, and some strains lack the ability to make them. Disease-causing bacteria with capsules such as *S. pneumoniae* and meningococci are better able to start an infection because the capsules make it difficult for white blood cells and antibodies to bind to the bacteria. The capsule may also buffer the effects of changing water content in a bacterium's environment. Capsules help groups of bacteria clump together. Some species are grown commercially for the gummy capsule material, which may include cellulose, glucan and dextran, which has a number of uses.

The **cell wall** is a tough material, murein, with subunits of peptidoglycan. It forms a meshwork coarse enough for even large enzymes and fragments of DNA to pass through. Dr Hans Christian Gram developed a staining technique for bacteria in 1884. Bacteria are described as **Gram-positive** if they take up the stain and **Gram-negative** if they do not. The response to the stain divides bacteria into two groups whose cell walls are different. Gram-positive bacteria have thick cell walls with over 40% peptidoglycan, together with other molecules, many of which provoke an immune response during infections. **Penicillin**, an antibiotic, affects

Gram-positive bacteria such as staphylococci and bacilli because it interferes with the synthesis of peptidoglycan in new bacterial cells. The molecules in such cells are not cross-linked properly so the wall is weaker than usual and osmotic stresses burst, or **lyse**, the newly formed cells.

Gram-negative bacteria have a thinner but more complex two-layered cell wall. There is little peptidoglycan and the outer layer is a membrane with proteins, lipids and lipopolysaccharides. Gram-negative bacteria, for example *E. coli*, are not susceptible to penicillin.

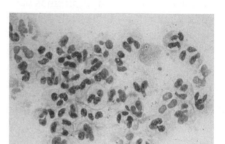

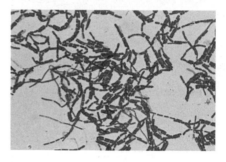

FIG 2.5 Gram-positive bacteria take up the stain and appear dark; Gram-negative bacteria appear lighter as they are stained with a different pink stain.

SPOTLIGHT

Gram's stain

A smear is made of a sample of actively growing bacterial cells on a slide then heat fixed. The cells are stained with crystal violet followed by dilute iodine. The slide is washed off with a mixture of ethanol and propanone. Gram-positive bacteria retain the stain in the meshwork of the cell wall, giving them a violet-purple colour, but Gram-negative bacteria do not. The Gram-negative bacteria are counterstained with safranin, or other stains, which stain them pink. Figure 2.5 shows Gram-negative and Gram-positive bacteria on two separate slides.

Some Gram-negative bacteria have rigid cylindrical rods made of protein projecting from their walls, called **pili** or fimbriae. These allow the bacteria to adhere to a surface, and to link to each other. A sex pilus is a longer projection that allows a cell to donate nucleic acid to another cell. You can read about this process in the next section.

Few bacterial species can move independently; they move about randomly in air currents and moisture films. There are some species with **flagella** that can swim through moisture films or water. A bacterial flagellum comprises a protein fibre starting from the cell membrane that spins round on its axis and propels the bacterium along. Bacteria may have one flagellum, like *Vibrio cholerae*; a tuft of a few, like *Helicobacter pylori*, involved in stomach ulcers; or have them all over their surface like *Salmonella* and *E. coli*. Where there is more than one flagellum they all rotate together, and may be reversed together.

The interior

Inside the cell wall is a plasma membrane, similar to that in animal and plant cells. In some places it is extended, for example into thylakoids in photosynthetic bacteria, and into folds called **mesosomes** near the site of cell division. Folding gives the membrane a larger surface area, which is important since it is the site of most metabolic activity in the cell, and the more surface the more activity there can be. The membrane carries enzymes for ATP-generating systems and for synthesising cell materials, as well as controlling the import and export of materials. Pigments and enzymes for photosynthesis are also located on this membrane. In photosynthetic bacteria the enzymes involved in carbon dioxide fixation may be found in organised folds called carboxysomes.

The interior of the cell is filled with **cytoplasm**. There is no separate nucleus but nucleic acids are found in a region of denser cytoplasm in the centre of the cell. Bacterial ribosomes are smaller (described as 70S) than those in animal cells and found scattered through the cytoplasm. Polysomes are groups of ribosomes near the membrane. Materials for growth and maintenance are spread through the cytoplasm, which also contains inclusions. These may be storage chemicals, for example glycogen granules, polyhydroxybutyric acid and lipid droplets, in varying amounts according to the environment and the cell's age. Some bacteria have inclusions concerned with CO_2 fixation or containing magnetite, which may help bacteria orientate themselves in a magnetic field.

Endospores are survival structures produced in adverse conditions by the bacilli and clostridia. An endospore develops inside a cell and is released when the cell is ruptured. Spores are extremely resistant and can remain viable for many years, withstanding the effects of drought, ultraviolet radiation, heat and many other stresses. The spore's position within the cell is used as an identification feature for these groups.

QUESTIONS

2.1   How is Gram's stain useful in the study of bacteria?

2.2   Explain why some bacteria are affected by penicillin but others are not?

2.3   How do bacterial cell walls differ from the cell walls of flowering plants?

2.4   Copy Table 2.1 when you have read the section about bacteria and complete the table using examples mentioned in the text.

TABLE 2.1 Gram-positive and Gram-negative bacteria

GRAM-POSITIVE	GRAM-NEGATIVE
Listeria monocytogenes	Escherichia coli
Staphylococcus aureus	Salmonella enteriditis
Lactobacillus	Clostridium spp
	Azotobacter

2.5   Where, and why, do bacteria make endospores?

DNA, reproduction and genetic exchange

Bacteria have a circular molecule of DNA, or **chromosome**, millions of base pairs long, certainly large enough in *E. coli* to encode potentially two or three thousand genes. The DNA base sequences of many bacteria have been deciphered but the functions of most of the genes have yet to be established. As bacteria have only one copy of each gene, any mutations are expressed straight away, and this promotes rapid evolution of new features and capabilities.

Bacteria reproduce asexually, by **binary fission.** When the cell divides, duplicated DNA is separated and the cell membrane folds inwards to form a double layer across the long axis of the cell. New cell wall layers are secreted within the membrane layers. This divides the cell into two smaller identical cells that may remain together or they may separate; dividing cells are shown in Figure 2.6. Binary fission is extremely rapid and will continue as long as the environment is favourable, allowing bacteria to exploit niches quickly.

DNA duplication is not linked directly to cell division so some bacteria may have more than one complement of DNA. DNA duplication starts with the DNA molecule attached to the cell membrane. Copying starts at this origin and continues round the molecule until it returns to the beginning.

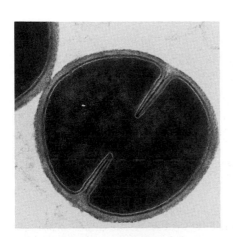

FIG 2.6 The two daughter cells made in binary fission become separated by a newly synthesised cell wall.

Bacteria do not exchange genetic material when they reproduce but they can acquire new or different genes from each other in other ways. This is sometimes referred to as **horizontal gene transfer** between species. For example, a cholera outbreak in India in 1969 was found to be caused by a strain of *E. coli* which had acquired a gene from the cholera-causing *Vibrio cholerae*. If the new DNA survives the recipient cell's DNAses, it may be incorporated into the genome. Bacteria that have gained new genes are said to be transformed. There are three main ways of gaining genes.

■ **Transformation**: Some bacteria gain genes from their environment. Extra DNA duplicated within a bacterium can accumulate outside it, in the capsule. This DNA quickly breaks up into smaller fragments that are absorbed by other bacteria and may be incorporated into their genome.

■ **Transduction**: Bacteria can also gain new DNA when particular viruses infect them. These viruses have reproduced inside other bacteria and carry fragments of bacterial DNA incorporated in their particles (you can find out more about these viruses in Section 2.6).

■ **Plasmid transfer**: About 70% of bacteria, both Gram-positive and Gram-negative, carry **plasmids**, which are a very important source of genetic material. Plasmids are small self-replicating loops of DNA which are separate from the main chromosome. They are about 1% of the size of the bacterial chromosome, up to about 200 kilobase pairs. Plasmids are not essential for bacterial metabolism, but often carry useful genes that enhance the survival of the bacteria carrying them. For example, in infection-causing bacteria they include drug resistance genes and an ability to infect more effectively. Plasmids are transferred fairly freely between bacteria of the same and related species. The transferred plasmid is duplicated or may be incorporated into the new bacterium's chromosome. Some bacteria, for example *E. coli*, carry an F plasmid that allows direct genetic exchange between two cells through sex pili. The F plasmid passes from the donor through the pilus to the recipient and takes some copies of donor genes with it. This process is called **conjugation**.

Plasmids are useful for genetic modification processes because they can be used to take genes, even those not normally found in bacteria, into particularly useful host bacteria. The transferred genes will be duplicated and used by the new host bacterium. These transformed bacteria are used to make useful products, such as human growth hormone. You can read more about these processes in Chapter 6.

QUESTIONS

2.6 Why do you think biologists refer to the size of DNA in 'kilobase pairs' rather than Relative Molecular Mass?

2.7 Explain what a plasmid is.

2.8 Sexual reproduction involving two individuals does not occur in bacteria. Outline how bacteria carry out genetic exchange.

Nutrition

Bacteria can carry out a wide range of biochemical activities, many of which are unique to the group. Collectively they have more ways of gaining energy and the materials they need for growth than any other type of organism. Like animal and plant cells they need an energy source and carbon-containing compounds, but some use materials that no other organisms can, and some synthesise compounds unique to bacteria. It is said that all naturally occurring carbon compounds can be metabolised by bacteria. Different bacterial species may obtain their energy from

- light,
- the oxidation of organic compounds such as sugars,
- oxidation of inorganic compounds such as hydrogen sulphide,
- geothermal energy from volcanic vents on the deep sea floor – though little is known about these species yet.

Bacteria can be grouped according to how they gain energy.

Photoautotrophs such as green and purple bacteria use light energy to fix carbon dioxide in photosynthesis. They release sulphur, not oxygen, because they use compounds such as hydrogen sulphide in the process instead of water. The sulphur may form deposits or may be oxidised to sulphate ions.

Green bacteria are confined to oxygen-free environments. They are found in the deeper parts of shallow ponds with other bacteria that also thrive in the low oxygen conditions. The purple and green bacteria use bacteriochlorophyll and carotenoids to harvest longer light wavelengths that have not been absorbed by plants and algae in the surface waters.

Chemoautotrophs use energy released by inorganic oxidations to drive ATP production and the synthesis of organic molecules from carbon dioxide. They are widespread in soils and water and include the nitrifying bacteria in the soil. Energy sources include hydrogen, nitrite ions, iron(II) compounds, and hydrogen sulphide. Many of these substances are the products of other bacterial activity and from geochemical reactions. These reactions are being harnessed to extract valuable metals such as niobium – more in Chapter 5. Chemoautotrophs can be grown in the dark in a mineral salts solution using carbon dioxide from the air, but some species need extra growth factors such as vitamins to grow.

Heterotrophs are the most common type. They use a wide range of organic carbon compounds, as shown in Table 2.2. No matter how complex the nutrient molecule, it is degraded into a form that can enter the main pathways of respiration, eventually being oxidised and the energy used to make ATP in a way similar to animals and plants. You can find out more about bacterial nutritional needs in Section 3.4 – Growing Microorganisms.

TABLE 2.2 Some carbon-containing compounds metabolised by microorganisms

alcohols
cellulose
chitin
hydrocarbons
methane
purines
steroids
waxes
polyethylene
pyrimidines
polyurethane
monocarboxylic acids

Respiration

Respiration is the release of energy from nutrient molecules. Animal and plant cells generally use oxygen in the process and release carbon dioxide. Bacteria use more varied pathways. Some bacteria are **aerobic**, using oxygen in respiration and releasing carbon dioxide. **Anaerobic** respirers, such as *Bacteroides fragilis*, use other molecules, such as nitrate and sulphate ions, instead of oxygen. Oxygen may inhibit their growth or may even be toxic. **Fermenting bacteria** break down glucose and other compounds to pyruvate conventionally, but then pyruvate is metabolised in different ways, again without using oxygen. These bacteria generate lactic acid and other organic acids, ethanol or a number of other compounds. Fermenters can grow easily in a wide variety of low oxygen habitats such as mud, animal gut and dead tissues. Lactic acid producing fermenting bacteria are widely used in the food industry (see Chapter 4).

Some bacteria normally use aerobic respiration but can ferment if there is not enough oxygen. These are known as **facultative anaerobes**. They include *E. coli* but the best known facultative anaerobe is in fact not a bacterium but a fungus, *Saccharomyces cerevisiae* (brewer's yeast). An organism which can only use one type of pathway is described as **obligate,** for example *Clostridium acetobutylicum* can only use anaerobic respiration.

SPOTLIGHT

Escherichia coli – a typical bacterium

This well-known bacterium is a member of the enteric group of bacteria, which inhabit the intestines of many animals and humans. It is Gram-negative with small rod-shaped cells. It can move about through moisture films using the flagella distributed all over the surface of the cell wall. Like other members of this group it can ferment sugars in low-oxygen environments such as animal gut but can use a wide range of nutrients in aerobic conditions. It synthesises glycogen as a nutrient store. It reproduces rapidly by binary fission, in less than 15 minutes in the most favourable conditions, but does not make spores. *E. coli* strains carry plasmids and can donate copies to other bacteria via specialised pili. *E. coli* is one of the organisms whose entire genome is being sequenced. It does not usually cause major disease problems, but strain *E. coli* O157 has become important; you can read more in Chapter 4.

Bacterial nutrition and environmental cycles

Bacteria are responsible for much of the breakdown of organic material within ecosystems, recycling materials and releasing nutrients that then become available to other organisms. Bacteria play a major role in the cycling of carbon, nitrogen and sulphur through ecosystems. Many of the naturally occurring pure iron ore and sulphur deposits are thought to be due to microbial activity.

Nitrogen-fixing bacteria can use atmospheric nitrogen when other sources, such as ammonium ions, are scarce. Nitrogen is reduced by the nitrogenase enzyme to ammonia, which is then used to make amino acids and other nitrogenous compounds. Most phototrophic bacteria and some other free-living Gram-negative bacteria, such as *Azoterbacter* and clostridia, can fix up to 3 kg of available nitrogen per hectare per year. There are symbiotic species such as *Rhizobium* that can fix very much more. These bacteria can live free in the soil but if a suitable host plant grows nearby, in this case members of the pea and bean family, they multiply and enter through young root hairs. Inside the plant they stimulate growth hormones, which cause cells to multiply and make nodules on the roots. A nitrogen-fixing gene has been transferred from one species to another, raising the possibility of engineering nitrogen fixation into crop plants.

Proteins and other nitrogen-containing compounds in dead animal and plant remains are broken down in decomposition to ammonia. This is oxidised by *Nitrosomonas* to nitrite, which is then oxidised by *Nitrobacter* to nitrate. Both these organisms grow poorly in low-oxygen, acidic, water-logged soils, thus reducing the fertility.

Anaerobic respiration by bacteria such as *Pseudomonas*, and *Thiobacillus denitrificans* activity, depletes the supply of nitrate ions in soil, releasing N_2 gas. Anaerobic bacteria play an important role in the rumen of animals such as cows, where they break down carbohydrates to small fatty acid molecules used by the ruminants for energy. Often, microorganisms use the products from the respiration of other organisms in the same ecosystem, forming a mutually interdependent community.

2.9 Use your biology course textbook to write out a definition of the following terms: autotroph, heterotroph, nitrogenous compound.

2.10 What is meant by the term 'chemoautotrophic nutrition' in bacteria? Are chemoautotrophs found in other groups of living organisms?

2.11 Reread the paragraphs on bacterial nutrition. What do you think is meant by the terms 'strict', 'facultative' and 'obligate'?

2.12 Figure 2.7 shows light absorption by pigments in bacteria and plants. Explain why green bacteria can photosynthesise deeper in the pond than plants can.

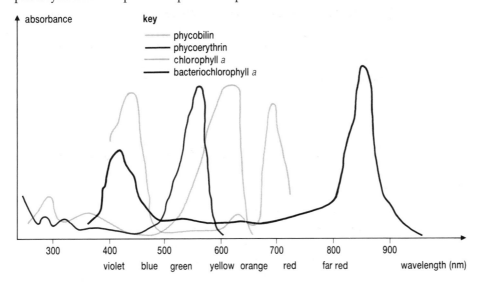

FIG 2.7 Absorption spectra of microbial pigments.

2.13 Reread the section above and locate where in the nitrogen cycle the following bacteria are active: *Nitrosomonas* which oxidises ammonium ions and *Nitrobacter* which oxidises nitrite ions. Draw a diagram to represent the nitrogen cycle and label the processes where these bacteria are active. You may need to refer to your main course textbook for the stages of the cycle.

2.14 For further reading: investigate the role played by autotrophic organisms in the cycling of nitrogen and sulphur.

2.3 CYANOBACTERIA

Cyanobacteria, or blue-green bacteria, are similar to photosynthetic Gram-negative bacteria. The cells contain a range of photosynthetic pigments that result in varied colours, which give the group its name.

Ecology and importance

Cyanobacteria are important photosynthesisers in aquatic ecosystems and are responsible for much of the productivity. Their photosynthesis fixes large amounts of CO_2 and they can also fix nitrogen when there is not enough suitable combined nitrogen. They grow in nutrient-poor waters and soils as long as the temperature and light are adequate. Some species are unicellular and float freely or buoyed up by gas vacuoles in surface waters, but others are linked in filamentous chains of cells attached to plants or solid surfaces in water or moist places. For many years they were thought to be algae because of the filaments, and were referred to as blue-green algae, but details of cell structure showed they were prokaryotic cells; nevertheless the term is still in everyday use. *Spirulina* is a filamentous species that grows well in salty alkaline lakes in Mexico and Chad, where people harvest it and dry it in the sun to make a protein-rich food. Each cell can lead an independent existence and reproduce when the chain of cells is fragmented.

When conditions are favourable cyanobacteria thrive in large numbers, colouring the water in an 'algal bloom'. Some species, such as *Microcystis*, release a toxin into the water that adversely affects animals coming into contact with it.

Structure and metabolism

Internally cyanobacteria are exceptional among prokaryotes because they have photosynthetic pigments in a series of coiled membranes, or **lamellae**. All cyanobacteria contain chlorophyll *a* and other pigments which allow them to harvest the greenish-blue light in the range 550–700 nm in deeper water. Nitrogen shortage involves about 10% of cells becoming **heterocysts** specialised for nitrogen fixation.

Cyanobacteria reproduce asexually; some bud off new cells from older cells, others reproduce by binary fission. Filaments may fragment into shorter lengths, followed by binary fission within the smaller chains.

QUESTIONS

2.15 Why are blue-green bacteria now grouped with bacteria instead of algae?

2.16 Give two reasons why blue-green bacteria are ecologically important.

2.17 Why are people advised not to carry out water sports in lakes with an 'algal bloom'?

2.4 PROTOCTISTA

The **Protoctista** kingdom includes Protozoa, the Euglenoids and the Chlorophyta or green algae. The protoctists are eukaryotic and unicellular, though some algae live as colonial forms. Each group has distinguishing characteristics but some microorganisms have features belonging to more than one group and so are hard to assign.

Protozoa

Protozoa live in moist habitats, such as soil, ponds and rivers, the sea and animal body fluids, where they tolerate a wide range of oxygen concentrations and pH. Many are herbivorous or detritus feeders, playing a large part in aquatic and soil food webs; others are predatory, feeding on bacteria and other protozoa. A few highly specialised protozoa are parasitic, causing important human and animal diseases including malaria, sleeping sickness and amoebic dysentery. This section gives only a brief survey of protozoal biology but you can read more about malaria and the protozoan that causes it in Chapter 8.

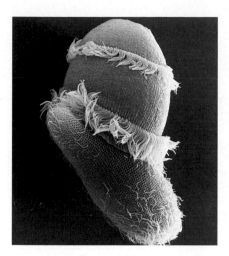

FIG 2.8 Both *Didinium* (the predator) and *Paramecium* (the prey) move by the coordinated action of rows of hair-like cilia.

Protozoa are relatively large for microorganisms, up to 5 µm across, and big enough to see with a good optical microscope. Protozoa have a tough cell membrane, or **pellicle**, but no cell wall. Most can move independently. One group has a characteristic crawling movement, for example *Amoeba proteus* moves along a surface by extending a portion of cell cytoplasm and membrane; the rest of the cell is gradually drawn forward in the same direction. Similar extensions flow round food particles, enclosing them in a food vacuole where enzymes break down materials; anything that cannot be absorbed is voided as the organism moves on.

Trypanosomas, a parasitic protozoan that causes human sleeping sickness, moves by using a single flagellum that is attached along the length of the cell and projects beyond. As the flagellum moves, it moves the attaching membrane and propels the organism through the blood. Other protozoa move using rows of cilia beating in a coordinated way to propel themselves through water films (Figure 2.8). Freshwater protozoa gain water as a result of osmosis. The animal uses respiratory energy to collect surplus water into a **contractile vacuole**. When it is filled it discharges at the cell surface. If conditions become too harsh, some protozoans can form dormant cysts.

Reproduction

Protozoa mainly reproduce by binary fission and can generate a large population quickly in favourable surroundings. In binary fission the cell nuclei and organelles are duplicated and are shared, with the cytoplasm, between two daughter cells. The process is quick, generating large populations rapidly. Some parasitic species undergo multiple fission that generates very large populations. In adverse conditions genetic exchange takes place during sexual reproduction, usually involving the fusion of pairs of similar gametes.

Unicellular green algae

Unicellular green algae, or Chlorophyta, are found in freshwater and seawater as plankton. They are the most common photosynthesisers in aquatic ecosystems and contribute greatly to productivity. They are also found in soil and moist environments; the green encrustation of *Pleurococcus* is a familiar sight on the bark of trees and damp walls. Many tolerate extreme habitats; species can even be found living in melted snow. Some species form mutualistic relationships with other organisms such as sponges, protozoa and fungi, where each partner benefits from the activity of the other. **Lichens** are mutualistic relationships between fungi and algae.

Calvin used a unicellular green alga, *Chlorella*, in his investigations into the processes of carbon fixation in photosynthesis. *Chlorella pyrenoidosa* is grown as a health food supplement, and strains are also grown to host viruses used as a source of enzymes in DNA research and modification. Potentially, unicellular green algae could be grown as sources of single-cell protein, vitamins and useful biochemicals. They can take up metal ions from low concentrations in surrounding water and this could be exploited to purify contaminated water or to extract precious metals from wastes.

Structure and physiology

Unicellular green algae have cellulose cell walls, although desmids and diatoms make silicate shells instead. *Chlorella*, shown in Figure 2.9, is widely distributed and makes a thick green suspension in standing water in the summer. It is well understood because of its use in photosynthesis research. *Chlorella* has small rounded or oval cells up to 15 µm in diameter. There is a single chloroplast in the cell with a pyrenoid containing carotene.

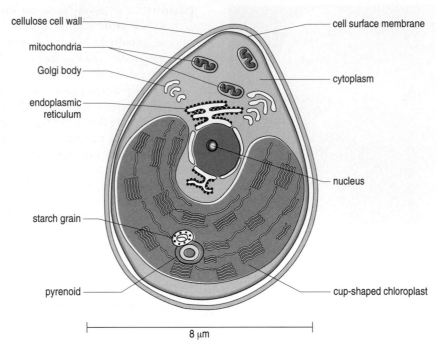

FIG 2.9 The structure of *Chlorella*.

Chlorella reproduces during darkness. Each cell produces four small daughter cells. Initially the daughters synthesise chlorophyll then synthesise other cell materials. The four daughters remain within the old cellulose wall for a while before they leave as independent cells.

SPOTLIGHT

Eutrophication

Many rivers, lakes and seas receive excessive amounts of nitrates and phosphates from time to time. One source is agricultural fertiliser because a substantial proportion of the average 150 kg of inorganic nitrate applied to each hectare of land each year is not taken up by plants, particularly if there is heavy rain soon after application. It is leached out or runs off the soil with rain and is carried as dissolved nitrate into watercourses or the water table. Another major source is discharges from sewage works which carry large quantities of these ions.

Blue-green bacteria and unicellular algae thrive in the nitrate- and phosphate-enriched waters, multiplying rapidly to reach huge numbers in the warmer surface water. These photosynthesising microorganisms are buoyed up by the oxygen they generate, forming a dense 'soup' that prevents light reaching plants in deeper water. When the microorganisms die they sink into deeper water and provide a rich source of nutrients for heterotrophic bacteria and other decomposers. These organisms deplete the available oxygen quickly and cause the death of many oxygen-dependent animals such as fish. Only anaerobic bacteria can grow in these conditions. Their respiration produces hydrogen sulphide and other substances which make the waters even more unsuitable for most aquatic organisms, and also make the water and mud smell very unpleasant. This process is called **eutrophication**.

QUESTIONS

2.18 How does the structure of a protozoan such as *Amoeba proteus* differ from that of a bacterium?

2.19 *Pleurococcus,* a unicellular green alga confined to moist places, often grows more densely on the northern side of a tree trunk. Outline an investigation you could carry out to determine whether moisture content of the air around the bark is an important factor in the distribution of *Pleurococcus.*

2.20 *Chlorella* is grown in Taiwan as a health food in large open-air ponds. What key checks or processes would the *Chlorella* have to undergo before it could be processed to make the food supplement?

2.21 Read the section on eutrophication and convert the process into a flow chart.

2.22 Research mutualism and Calvin's work in your course textbook.

2.5 FUNGI

Fungi are common in soil and water where they play a key role in the cycling of organic matter, releasing components that can be used by other organisms in an ecosystem. Fungi are among the first to degrade dead organic matter using a wide range of enzymes including cellulases and lignases. One group of fungi, the Mycorrhiza, live in mutualistic relationships with trees. Mycorrhizas enable pine and other trees to grow in nutritionally poor soils. The fungus grows around the roots of a specific host tree, taking sugars from its host; the fungus provides the tree with access to other nutrients such as nitrogen compounds.

Yeasts are used commercially for brewing and baking and other fungal species are grown to produce a range of useful enzymes and other substances such as antibiotics.

A large number of fungi are involved in plant diseases and there are some that infect animals. You can find out more about these in Chapters 8 and 9.

Structure

The least complex fungi are protozoa-like, and yeasts are single-celled, but most fungi are multicellular, seen as a widespread thread-like mass growing through a substrate. The thread-like mass is a **mycelium** made of multicellular filaments called **hyphae** that radiate through the substrate in which they are growing. Hyphae have walls composed of a strong polymer called chitin with a lining of cytoplasm and often a large vacuole. Many important fungal species, such as *Penicillium*, do not have the usual cell structure in hyphae; instead, the long tube of a hypha lacks cross-walls and is filled with a mass of cytoplasm with many nuclei spread through. The structure of a hypha can be seen in Figure 2.10. Older hyphae in the centre of the mycelium may no longer contain cytoplasm.

(a) hypha with cross-walls

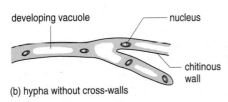

developing vacuole — nucleus

— chitinous wall

(b) hypha without cross-walls

FIG 2.10 The structure of hyphae. Even though these are eukaryotic cells, some species lack cross-walls separating the hypha into separate cells.

FIG 2.11 Fungal mycelia grow out in all directions, making a circular structure. The new hyphae at the edge mobilise nutrients and translocate them to older mycelium. They will also carry sporing bodies like these mushrooms in a 'fairy ring', showing the leading edge of the mycelium.

Each mycelium starts with the germination of a single spore that grows a hypha. The hypha branches, eventually forming a circular mat extending over the surface of the substrate. There is no limit to the size of the mycelium; it will grow wherever there are enough nutrients. The largest living organisms are fungal mycelia growing in the USA, the biggest thousands of metres across, covering 600 hectares. In large mycelia the oldest parts in the centre die, leaving a ring of younger actively growing mycelium. Figure 2.11 shows the spore-producing structures at the edge of a mycelium growing through grassland. When the immediate environment is exhausted many fungi can translocate nutrients from one part of the mycelium to another, supporting hyphae growing over stone or unfavourable substrates.

Metabolism

Fungi are strictly aerobic, and are usually confined to the surface layers of substances, where the oxygen concentration is high. Fungi are very tolerant of acidity and dryness and so can grow in substances with very low water potentials such as wood and jam, or as acid as tinned tomatoes and orange juice. Fungi obtain their nutrients from the action of enzymes secreted through the hyphal walls. They degrade organic material then absorb soluble products back through the cell wall. Collectively fungi produce a very wide range of enzymes, including some found in few other organisms, for example cellulases and lignases, which make them important in the cycling of carbon through ecosystems.

Reproduction

All fungi can grow from fragments of mycelium, which will grow into a new mycelium. They can also produce large numbers of sexual and asexual spores from sporing structures. In favourable conditions the mycelium grows rapidly and produces asexual spores within a few days. Sexual spores are produced when the environment becomes harsher. Spores are released into the air or water and disperse widely, some travelling hundreds of kilometres. Warm moist conditions are ideal for their germination.

Zygomycetes such as *Rhizopus*, or black bread mould, and *Mucor*, or pin mould, make sexual spores in chambers called sporangia, at the tips of hyphae. These are easy to see in terrestrial species as they project above the mycelium. Some species such as *Mucor* have different mating strains but it is difficult to distinguish any physical differences between them.

Ascomycetes are common and commercially important fungi. As well as yeasts the group includes *Penicillium* which makes penicillin, *Aspergillus* which causes lung problems, powdery mildews on crops, morels and the truffle, an extremely expensive food ingredient. They cause many problems, from the rotting of cloth to apple scab, blue, green and black moulds, ringworm, wheat diseases and Dutch elm disease. You can find out more about the commercial use of yeast and other fungi in Chapter 4.

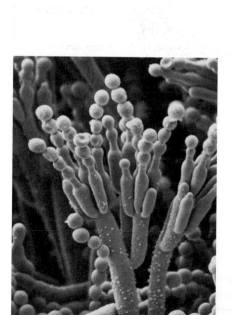

FIG 2.12 Chains of asexual conidiospores develop on the tip of aerial hyphae (here *Penicillium* sp) for easy dispersal in air currents.

Ascomycetes produce plentiful asexual **conidiospores** from specialised hyphae during active growth; some can be seen in Figure 2.12. The long chains of spores are often brightly coloured and colour the mycelium before being dispersed in air currents. You may be familiar with the green of *Penicillium* species growing on citrus fruit. The sexual spores are **ascospores**, produced in a different structure.

Yeasts are different to the rest of the Ascomycetes. The many yeast species do not form mycelia but exist as single cells reproducing by budding. Yeasts are described in more detail in the Spotlight section.

The **basidiomycetes** make the familiar large fruiting bodies called mushrooms and toadstools, though the fruiting body may be microscopically small in many species. The group includes bracket fungi, bird's nest fungi, smuts and rusts. Most of the year the fungus grows through its substrate – soil, wood or vegetation – as an extensive mycelium. Basidiomycetes seldom make asexual spores, but produce sexual **basidiospores** in the fruiting body.

Yeasts

Yeasts are found everywhere – on the skins of fruit and vegetables, in dust, dung, soil, water, milk, and on animal mucous membranes. They have been used since prehistoric times to ferment fruit and grain sugars to make alcoholic drinks and as leaveners in bread. The term 'yeast' is really a non-specific name for a whole group of organisms which can be divided into two groups. One group uses aerobic respiration and does not make ethanol. The presence of oxygen stimulates the breakdown of pyruvate, completely oxidising sugars to carbon dioxide. These yeasts grow rapidly and some, for example *Hansenula*, are important spoilage organisms. The members of the other group, which includes *Saccharomyces cerevisiae*, or baker's yeast, can tolerate low oxygen environments. They break down glucose to pyruvate to obtain energy then convert it to ethanol as long as there is plenty of free glucose in their environment. If the glucose levels fall, they switch to making ethanol into carbon dioxide. These products – carbon dioxide, which causes bread to rise, and ethanol which gives alcoholic drinks their potency – are the bases of vast industries. Surplus yeasts from brewing and baking are used as a source of Vitamin B complex, yeast extracts and savoury spreads, flavourings, and as animal feed, though increasingly these are being replaced by strains specifically bred for the purpose.

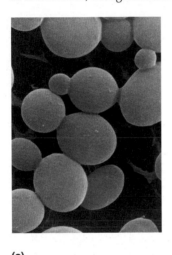
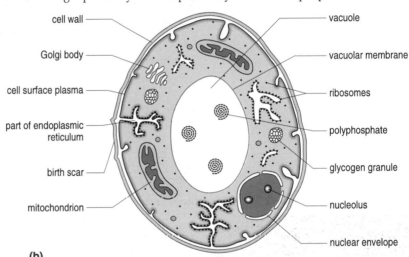

(a) **(b)**

FIG 2.13 (a) Unlike most fungi, yeast cells reproduce by budding smaller cells, which remain attached in clusters. **(b)** Sectional drawing of a yeast cell. Note the eukaryotic features.

Baker's yeast, *S. cerevisiae*, appears as ovoid cells surrounded by a thick cell wall. You can see a cluster of yeast cells in Figure 2.13. Some cells appear to have smaller cells attached to them. These are daughter cells produced by budding that have remained attached, forming clumps. Yeasts survive unsuitable conditions as dormant spores, produced from the fusion of two yeast cells.

There are hundreds of yeast strains, each with its own characteristics and abilities used for a range of products. Manufacturers are trying to improve their strains but it is difficult to breed improved varieties as yeasts do not mate randomly – only with cells of the same strain – making cross-breeding very difficult. Commercial strains are often genetically abnormal with multiple sets of chromosomes or with missing chromosomes. This too makes it difficult to do genetic crosses with them. The genome of *S. cerevisiae* has been sequenced and provided information about how the structural stability of the genome is maintained. However, we do not yet know the function of many of the genes yet. The information gleaned seems to be applicable to human genes too, so yeast looks like a good model system for work on human genes.

Research into improving yeast strains has a variety of targets, including strains that:
- ferment more quickly for quicker throughput,
- tolerate higher ethanol concentrations for stronger drinks,
- produce less alcohol with the same flavours for low-alcohol drinks.

Efforts are being made to engineer desirable genes for entirely new products into yeast cells, such as *S. cerevisiae*, *Hansenula* and *Kluyveromyces*. These are organisms whose growth requirements are well understood and could be used to synthesise a range of commercial biochemicals.

2.23 In industrial processes using *Aspergillus*, the inoculating material into the substrate is often described as a 'conidial suspension'. What do you think this means?

2.24 What is the importance of extracellular enzyme production in the life of a fungus?

2.25 Fungi readily grow on cotton fabric, which is over 90% cellulose, in warm, moist climates. If the fabric is tested, it is found to have lost much of its strength. Why do you think this loss of strength occurs?

2.26 What environmental factors limit the growth of a fungal mycelium?

2.27 If your house is found to suffer a dry rot fungus infection (*Serpula lachrymans*) you are recommended to remove infected wood and surrounding wood and to burn both. Why does the wood around the infection have to be burned as well?

2.28 Read through the section above and explain what determines when a fungus makes sexual or asexual spores.

2.6 VIRUSES

Viruses are very different from any of the previous groups, as they do not have a cellular structure. Outside cells they are inert particles called **virions**, made of a protein coat wrapped round a nucleic acid molecule but with no other cell components. Virions can infect susceptible cells but have no other abilities. Only when they gain entry to a living cell can they show properties of living organisms. Inside susceptible cells they usurp normal cell controls and divert cell activity into virus-directed activity. This disruption of cell activity usually has harmful effects on the host. There are a few temperate viruses which do not cause damage immediately. These lodge in cells, lying latent, often inserted into the chromosomes, and are duplicated with the cells.

Viruses are specialised for infecting a specific host, and there are virus parasites of almost all living organisms. Viruses that infect bacteria are called **bacteriophages**, or phage for short. Viruses are grouped by their structure and the type of genetic material. An increasing understanding of viruses has given us insight into the fundamental genetic processes occurring in cells, and has led to some of the techniques used in genetic engineering.

Structure

Viruses are very small, ranging from about 20 nm to 400 nm in size. Each virus has a simple structure, namely a molecule of nucleic acid as its core, arranged within a protein coat, or **capsid**. The capsid is made up of protein subunits called **capsomeres**, which link together spontaneously in regular arrays to form geometrical structures. The most common shapes are twenty-sided (icosahedral) capsids, and helical capsids. There may be more than one type of protein making up the capsid. Some proteins specifically bind to receptors on the host cell, which accounts for the need for a specific type of host. Others help the virus to enter the cell. Figure 2.14 shows the structures of some interesting viruses.

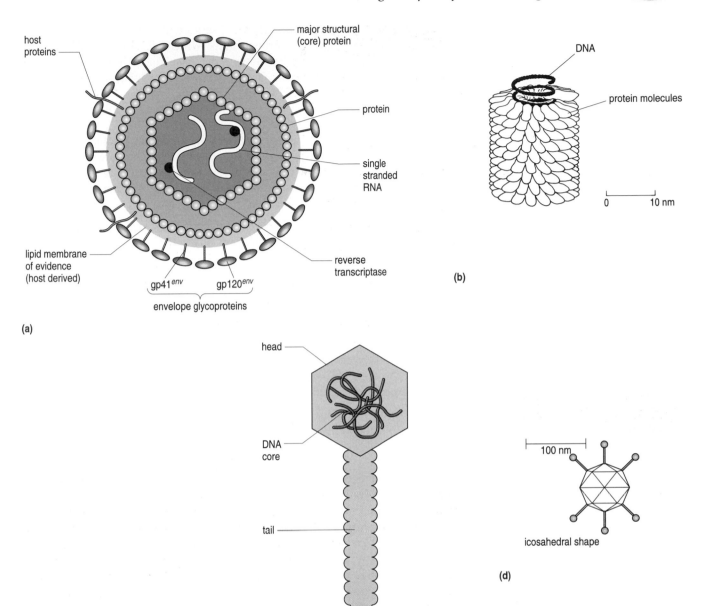

host proteins

major structural (core) protein

DNA

protein molecules

protein

single stranded RNA

lipid membrane of evidence (host derived)

reverse transcriptase

gp41*env*　gp120*env*

envelope glycoproteins

(a)

(b)

head

DNA core

tail

(c)

100 nm

icosahedral shape

(d)

FIG 2.14 (a) HIV is an icosahedral RNA virus; **(b)** TMV is a helical DNA virus; **(c)** lambda phage is a DNA bacteriophage that infects *E. coli* bacteria; **(d)** adenovirus is a DNA virus. It infects human respiratory cells and is used to carry genes into human cells.

Viral genes are carried as a molecule of DNA or RNA coiled within the capsid. Some viruses have double-stranded DNA, like other organisms, but others, such as influenza and HIV, have RNA as a single-stranded or even double-stranded molecule. There are even single-stranded DNA viruses. The genome of the smallest virus, øX174, is about 10 genes but the nucleic acid in larger viruses is long enough to encode several hundred genes. A few viruses, like influenza, acquire a membrane layer called an **envelope** around the outside of the capsid as they emerge through the host cell membrane. Figure 2.15 shows enveloped viruses emerging from a cell.

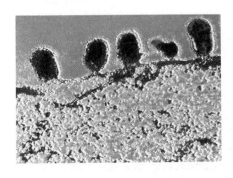

FIG 2.15 Influenza virus particles emerge from a host cell and acquire a coating of membrane during their passage.

Some bacteriophages, particularly the T series and lambda (λ) phage, have a complex structure. They have protein 'tails' which are able to combine with receptors on the host cell surface, allowing the virus to attach or adsorb onto its host.

Life cycle

The life cycle of any virus is affected by its lack of independent mobility. Many, like the common cold viruses, rely on passive dispersal as they are moved about randomly in fluids or air, and can only infect when they encounter a suitable cell. However, some animal and most plant viruses are transmitted from one host to the next by a vector.

The cycle of infection begins with one or more viruses encountering a suitable host cell and entering. The entry method depends on the type of cell being infected.

■ Animal viruses adsorb onto chemical receptors in the host cell membrane and are then taken through it by a process similar to endocytosis.

■ Bacteriophages such as λ in Figure 2.14 use the tail proteins to attach to the host bacterium's wall, then another contractile protein in the tail assembly injects the nucleic acids into the host cell, leaving the capsule outside.

■ Plant viruses cannot enter tough plant tissue by themselves, unless the cells are damaged. Vectors such as greenfly feeding on plant cells inject them. Once inside, the viruses can pass from cell to cell through plant transport tissues. You can read more about this in Chapter 9.

Viruses are only infectious as whole particles; without the coat they cannot attach to infect. Animal and plant viruses are uncoated as part of the entry process and only nucleic acid can be detected inside the cell. At this **eclipse** phase there are no virus particles capable of causing further infection, and so virus numbers are harder to count. Nevertheless the virus nucleic acid is active. Viral nucleic acid molecules have two functions: they act as templates for the synthesis of more nucleic acid for packaging into progeny viruses, and they also carry genes for viral proteins needed for the synthesis and structure of progeny particles. Some of the viruses with single-stranded RNA are 'read off' or **translated** by the host cell ribosomes directly; others may need a complementary strand of RNA or DNA synthesised first. Proteins, including those needed for duplicating the nucleic acid, are synthesised, though some DNA viruses may have replicating enzymes enclosed in the capsid or may be able to use host enzymes for duplicating and transcribing their DNA. Normal host cell metabolism and DNA activity is often severely disrupted by the activity of the virus.

As viral nucleic acid and proteins accumulate within the cell, capsid proteins take on their three-dimensional forms and the capsid assembles spontaneously. The nucleic acid is packaged, and then the viruses leave the cell. Some are liberated by cell membrane rupture, or they may pass through the host cell membrane, picking up an envelope on the way. The pattern of events is often known as the lytic cycle, as it may end in cell rupturing, or **lysis**, to release the viruses. You can read more about the life cycle and diseases caused by 'flu and HIV in Chapter 8.

Lysogenic bacteriophages, or temperate viruses, are bacteriophages which, when they invade the cell, do not undergo the full infective cycle but are found only as DNA and are known as prophages. They are duplicated with host DNA and most of their genes are repressed. λ phage integrates into the chromosome of *E. coli* strain K12 bacteria. When the bacteria divide by binary fission a copy of λ goes in the DNA into daughter cells. Other prophages may exist as plasmids. These viruses can be induced to replicate by stresses such as ultraviolet light or by biochemical changes in the cell. In some bacteria, for example the diphtheria bacteria, the presence of certain prophages enables them to produce a toxin that they could not otherwise make.

SPOTLIGHT

Retroviruses

Retroviruses are RNA viruses that carry an enzyme in their capsules called **reverse transcriptase**. Reverse transcriptase can synthesise a strand of DNA using its RNA molecule as a template. The virus can then synthesise a complementary strand of DNA to form a circular chromosome. This chromosome migrates to the nucleus of the host cell and is inserted into the host cell DNA. It is then used for the synthesis of viral proteins like any other length of host genome. The enzyme is widely used in genetic modification techniques.

The group has become more widely known as one of its members, HIV, shown in Figure 2.14, has been established as the cause of Acquired Immune Deficiency Syndrome, usually referred to as AIDS. You can read about the HIV infection in Chapter 8. The retroviruses are scientifically interesting because some of the group's members are among the few viruses linked directly with cancer. Also the action of reverse transcriptase conflicts with the idea of a one-way flow of genetic information from DNA to RNA to protein.

TABLE 2.3 Other interesting microorganisms

The **Mollicutes**: a bacterial sub-group including the mycoplasmas – the smallest cellular structures. They lack a rigid cell wall and so are flexible enough to squeeze through small gaps. Generally resistant to antibiotics. Many are fermenters found on mucous and synovial membranes where they cause many disorders. Others cause insect and plant diseases.

The **Chlamydias** are intracellular bacterial parasites. They do not seem to have ATP-generating systems of their own and are totally reliant on the host cell for their energy needs. One group causes psittacosis, a disease of birds that can cause lung disease in the handlers. Another group causes diseases such as trachoma, a major cause of blindness in underdeveloped countries, and infections of the urino-genital tract which are difficult to diagnose.

Viroids are RNA molecules without a capsid that can infect cells. They are implicated in several plant infections such as potato spindle tuber. The RNA is particularly resistant, heat- and UV-stable. It has been said that some viroid genomes are similar to some of the non-coding portions of eukaryotic genes which are spliced out before mRNA is released into the cytoplasm. It has been suggested that they may interfere with the normal cell mRNA editing process.

Euglenoids have both protozoal and algal features. They are single-celled, with a tough cell membrane and swim around using a flagellum. Most photosynthesise with chloroplasts but become heterotrophic in the dark, absorbing soluble organic nutrients from the surrounding medium. There are some colourless euglenoids which cannot photosynthesise and are therefore heterotrophic.

The **Rickettsias** are intracellular parasites. They seem to have a bacterial structure that can survive intracellular digestion. Most are animal parasites carried by lice or ticks; humans are infected accidentally. They cause diseases such as typhus and Rocky Mountain spotted fever.

The **slime moulds** can form a mass of multinucleate cells. The cells lack rigid walls and the cell mass is mobile, flowing in an amoeboid motion taking in particulate food. Colonies are fan shaped and may be very large. In dry substrates they may form complex fruiting bodies, releasing spores from which gametes develop. Gametes fuse to form a new slime mould. One sub-group is unique as they exist as protozoa-like uninucleate cells multiplying by binary fission. They can aggregate and cooperate to make a fruiting body producing asexual spores, yet each cell never loses its individuality.

Satellite viruses are small pieces of nucleic acid that need the aid of another virus to replicate. Many are very small RNA molecules which need proteins from a helper virus or host cell to complete their replication and subsequent infection of new cells. The activity of the helper virus may be slowed down, but satellites can also make helper virus infections worse.

Prions are infectious glycoprotein particles containing a polypeptide that cause diseases such as scrapie, CJD and BSE. Some seem to be misshapen forms of normal proteins.

QUESTIONS

2.29 Construct a flow chart of the life cycle of bacteriophage λ.

2.30 Is the antibiotic penicillin useful in treating virus infections?

2.31 Describe the structure of a named DNA virus.

2.32 Look up the work of Hershey & Chase, who used bacteriophages in their research work establishing whether genetic information is carried by DNA or protein (a popular idea at the time).

2.33 Draw up a table comparing the features of bacteria, fungi and viruses. Use headings such as size, mode of nutrition, structure.

2.34 Make a table of all the commercial uses of microorganisms mentioned in the chapter. Use the headings 'Organism' and 'Use'.

2.35 Describe how microorganisms can survive adverse conditions.

2.36 Describe the role of microorganisms in the cycling of matter through ecosystems.

2.37 Compare reproduction processes in microorganisms with those of mammals.

2.38 What role have microorganisms played in the development of genetic modification techniques?

2.39 Figure 2.16 shows the structure of a virus.

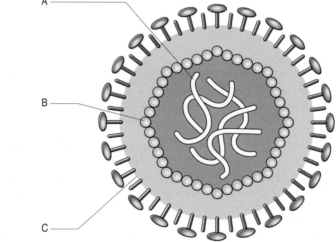

FIG 2.16

(a) Name the structures A, B and C.

(b) State two characteristics that are used to distinguish viruses from the other major groups of microorganisms.

(c) Outline the life cycle of a named bacteriophage.

Exam questions

2.40 Complete the following table. Use a tick to indicate that the characteristic is present or a cross to indicate that it is absent. (3)

Characteristic	Virus	Fungus	Bacterium
70S ribosomes			
mesosome			
mitochondria			
capsid			

AQA Biology, June 1999, Paper BY06, Q. 3, Abridged

2.41 The diagram shows a bacterium. Identify the structures labelled **A** to **F** and give **one** function of each. (6)

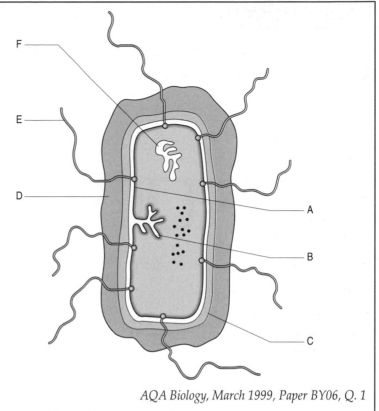

F

E

D

A

B

C

AQA Biology, March 1999, Paper BY06, Q. 1

SUMMARY

Microorganisms reproduce rapidly by asexual binary fission or by spores to colonise every ecological niche.

Many degrade organic material and play an important role in the decay cycle. Some bacteria gain energy from inorganic chemical reactions and are important geochemical agents.

Bacteria and blue-green bacteria are prokaryotic.

Gram's stain divides bacteria into two groups reflecting their cell wall composition.

Bacterial genetic exchange mechanisms have been exploited for genetic modification techniques.

Bacterial respiration may not require oxygen, and bacteria colonise oxygen-poor ecological niches.

Facultative anaerobes can switch to fermentation if there is insufficient oxygen for aerobic respiration.

Photoautotrophs use light energy to fix CO_2. Chemoautotrophs use chemical oxidations instead.

Plasmids are extra loops of DNA carrying genes that may be advantageous but are not essential, and which can be passed from one bacterium to another.

Bacteria do not reproduce by spore production; a spore is a survival structure.

Blue-green bacteria and green algae are the most important carbon fixers in aquatic ecosystems and a number of other habitats. Blue-green bacteria are important nitrogen fixers in nutrient-poor habitats.

Protozoa are fully independent eukaryotic unicellular animals. Most are detritus feeders or predate other microorganisms and are important in decay processes.

Fungi are multicellular heterotrophs and can degrade materials that other microorganisms cannot; they are among the initiators of the decay process. Fungi and green algae are eukaryotic.

There are parasitic bacteria, protozoa and fungi that cause disease in animals and plants.

Viruses do not have a cellular structure.

Viruses are parasites whose activity within cells usually disturbs host cell metabolism.

3 GROWING MICROORGANISMS

3.1 MICROORGANISMS IN THE BIOSPHERE

Microorganisms abound in all ecological niches, in freshwater and salt water, on land and in the lower atmosphere. They have been recovered from the depths of the Pacific Ocean where it is cold, dark and under high pressure, and from the stratosphere where there is little air, high UV levels and few nutrients.

Microorganisms are well adapted for exploiting a favourable environment. They spread widely because they are small, light and very easily dispersed by wind and water. They grow quickly because they have a rapid metabolic rate and their large surface area to volume ratio allows them to exchange materials easily with their environment. As they multiply they use materials immediately around them and excrete metabolites, gradually changing their environment. As they use up

FIG 3.1 The very pale streaks across the snow are caused by red algae – an organism which can survive harsh environments.

nutrients and release wastes they make the area less suitable for their own growth but often more suitable for another species, which will flourish in turn. In favourable conditions bacteria may reproduce as often as every 20–25 minutes. Most members of a microbial population die when conditions become very unfavourable, but a few survive in a dormant state, and so do spores. They grow rapidly when conditions become favourable again.

Warm, moist, nutrient-rich environments, such as animal mucous membranes and compost heaps, suit the needs of many microbial species, and competition will be intense. Some species tolerate a wide range of environmental factors, but others are specialised for a particular set of conditions, or just a few substrates. There are even species that thrive in harsh environments with several extreme physical factors such as Arctic waters or a frozen chicken. Each environment poses different problems for microorganisms, and they have evolved strategies that allow them to grow.

3.2 FACTORS AFFECTING MICROBIAL GROWTH

Temperature

Temperature is one of the most important environmental factors affecting microbial growth. Most species have a characteristic range of temperatures in which they can grow, but they don't grow at the same rate over the whole of their range. Growth is governed by the rate of chemical reactions within cells. Increasing the temperature leads to an increase in the rate of reactions; more materials are synthesised and growth is faster. There is an upper temperature limit for growth because the reactions are catalysed by enzymes, and heat affects the structure of enzymes and other proteins. Most proteins start to lose their three-dimensional shape at temperatures over 45 °C, their properties are altered and enzymes lose their activity. At 60 °C over half the proteins present in the cytoplasm of a nonadapted bacterium will have been denatured. Growth quickly stops when the temperature is high enough to denature proteins. Species adapted to hot environments have thermally stable proteins.

At the lower end of the temperature range we would expect growth to be slower, eventually ceasing when the cytoplasm freezes, but most organisms stop growing well before this. At low temperatures protein structure changes again. Cell membrane lipids also alter their physical nature so membrane functions are disturbed. Figure 3.2 illustrates the rate of growth of a typical bacterium over its temperature range. Notice that the lethal temperature is only a few degrees higher than an organism's **optimum temperature** – the temperature at which it grows most quickly.

FIG 3.2 How a bacterial species growth rate varies with temperature.

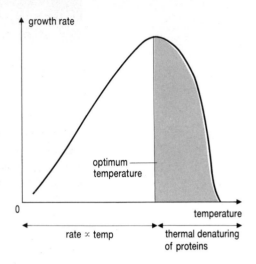

Most bacterial cells are killed quickly at 60 °C but some bacterial spores can endure temperatures over 100 °C. Table 3.1 shows the three overlapping groups into which microorganisms are placed according to their optimum temperature and range.

TABLE 3.1 Temperature range of some important bacteria

	PSYCHROPHILES	MESOPHILES	THERMOPHILES
growth	can grow at 0 °C though their optimum temperature may be slightly higher	grow in the mid-range of ambient temperatures and have optima between 20 and 40 °C	grow at high temperatures with optima over 45 °C
ecology	much of the ocean, the poles, and mountainous regions	ambient temperatures of most habitats includes most common bacteria, fungi, yeasts, viruses	sun heated rocks and soils, hot springs and power station cooling towers, rotting vegetation
significance	cause problems in frozen foods	recycling materials in habitats some human pathogens have a very narrow temperature range for growth	microbial activity in compost, slag and coal heaps, animal dung, and hay generates temperatures of over 60 °C source of heat-stable enzymes for use in biotechnological processes
examples temperature range	*Bacillus globisporus* −10 to +20 °C *Micrococcus cryophilus* −10 to +20 °C	*Escherichia coli* 5–40 °C *Lactobacillus delbrueckii* 17–50 °C *Neisseria gonorrhoeae* 30–40 °C	*Sulpholobus* 55–85 °C *Bacillus stearothermophilus* 35–70 °C *Bacillus coagulans* 25–65 °C

Oxygen concentration

The effect of variation in oxygen concentration depends on an organism's metabolism. Fungi, protozoa and many bacteria use oxygen for respiration. If the oxygen concentration falls, their activity declines accordingly. However, some species need only 2–10% oxygen for respiration. Their growth is not inhibited until oxygen declines to very low levels. **Microaerophiles** use oxygen but cannot tolerate concentrations as high as 20% and are inhibited at atmospheric concentrations. Many gut bacteria belong in this group. **Facultative anaerobes** use oxygen if it is available but ferment if it is not, so their activity does change. **Obligate anaerobes** do not use oxygen for respiration, and some, such as *Clostridium* species, are very sensitive to oxygen. Even very small quantities may be toxic as they lack the enzymes needed to detoxify peroxides made in aerobic reactions.

pH

Most organisms are sensitive to changes of pH in their environment. Most bacteria can grow between pH 4 and 9, but grow best at pH 6.5. The growth rate falls off quickly either side of the optimum. The presence of H^+ ions from acids affects the structure of proteins and other molecules. Most of the natural environment has a pH close to neutral but there are acid and alkaline environments. Bogs, soil in pine woods, some streams and mine waters are acidic whereas desert soils, animal dung, decaying protein and some lakes are alkaline environments. **Acidophiles** such as *Thiobacillus thiooxidans* will grow in extreme sites where the pH is 3 or less. Many acidophiles need high H^+ levels to maintain their structures. Fungi tolerate slightly more acidic environments than most bacteria.

Nutrient availability

All microorganisms need materials from their environments for growth. Photosynthetic bacteria and unicellular algae are affected by light intensity, which limits the rate of photosynthesis. Other bacteria need nutrients so their growth rate depends on the supply available. Bacteria multiply rapidly when nutrients are plentiful. Sections 3.4 and 3.9 contain more detailed information on how nutrient availability affects microbial growth.

Water potential

Very salty or sugary surroundings tend to draw water out of organisms by osmosis. Bacteria can tolerate a wide range of external solute concentrations because their rigid cell walls resist osmotic changes. Organisms adapted to the very high salt concentrations in salt lakes are known as **halophiles**. In fact, *Pediococcus* and *Halobacterium* can tolerate over 25% salt. Fungi also tolerate very low water potential, which is why they can grow in jam which is over 50% sugar. Species that can make dormant spores or cysts can survive desiccation but very few can grow where there is little free water available.

3.1   List the main factors that influence microbial growth.

3.2   In what sort of natural environment would you expect to find
(a) a halophile,
(b) a thermophile,
(c) an acidophile?

3.3   Which constituents of bacterial cells are most affected by changes in temperature? How are bacterial cells modified for life at high and low temperatures?

3.4   If you are not familiar with the term 'water potential' or the effects of temperature and pH on protein structures, you should read about osmosis or proteins in a biology textbook.

3.3   STERILE TECHNIQUE

Microorganisms are usually used as **pure cultures**, that is a culture containing only one species of microorganism. If other organisms get into the sample it is **contaminated**. Any investigations on, or product made by, that culture would be unacceptable as the contaminant's metabolism would have affected it. Contamination in an industrial process may result in losing a whole batch of product.

Keeping a culture free of contaminants requires systematic precautions and techniques. These techniques for handling cultures are called **aseptic techniques**, see Figure 3.3. The same procedures also stop the organisms escaping from their containers and contaminating the workplace or the people handling them. This is particularly important if the organism is harmful.

FIG 3.3 Aseptic techniques are used when sampling a culture to stop microorganisms either escaping from the culture or getting into it.

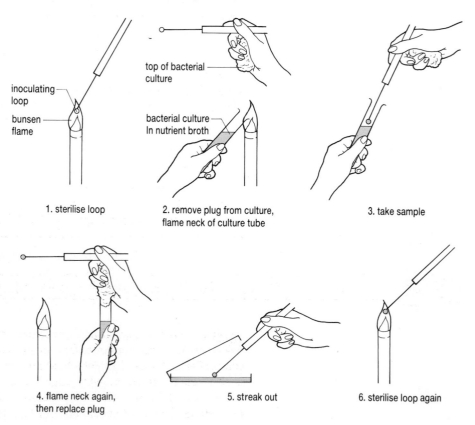

top of bacterial culture

inoculating loop

bunsen flame

bacterial culture In nutrient broth

1. sterilise loop

2. remove plug from culture, flame neck of culture tube

3. take sample

4. flame neck again, then replace plug

5. streak out

6. sterilise loop again

- Work surfaces are wiped with disinfectant before and after work.
- Everything, including equipment, instruments and materials, must be sterilised before and after use.
- The necks of containers are flamed when they are opened or closed, and lids are not left on benches or upside-down, which stops airborne organisms entering them.
- Anything sterile is not allowed to come into contact with unsterile materials.
- Particular care is taken with liquids containing microorganisms. A droplet on an inoculating loop that is heated quickly may form an **aerosol**, which spreads organisms into the air.
- Used equipment is placed in disinfectant before cleaning and resterilising.

Disinfectants such as sodium hypochlorite have a bleaching action that kills live cells. Glassware and metal equipment can be sterilised by placing in a sealed container and heating to high temperatures in an oven. Alternatively they can be sterilised in an **autoclave** (see Spotlight). Moist heat is far more effective at reducing the bacterial count than dry heat. Items adversely affected by heat are sterilised in other ways: for example, solutions can be passed through filters fine enough to trap microorganisms; or antibiotics may be added to prevent the growth of common bacteria. Rooms, smaller working areas and some equipment can be sterilised by bathing in ultraviolet light. Some working areas may have air flow hoods over them, which draw contaminated air from the workplace through filters – a bit like a fume cupboard.

Whenever anyone works with microorganisms a risk assessment must be carried out and appropriate measures taken. Containment is particularly important with dangerous organisms. These are usually handled within a transfer chamber, which is a special enclosed chamber, sometimes within a special room inside a laboratory. The air flow within the chamber draws air from the laboratory, and out through filters and traps within the hood. Only the user's arms, sometimes gloved for protection, enter the chamber. Even more measures are taken with very dangerous organisms. The whole laboratory complex has secure entrances and a negative air flow system where differences in air pressure draw air from outside through the laboratory and out through filters and traps, reducing the chance of organisms escaping when people pass through doors.

SPOTLIGHT

Autoclaving

Autoclaving is the process of heating items and solutions to high temperatures in a container filled with steam under pressure. This gives a temperature equivalent to 120 °C or greater. A domestic pressure cooker works in the same way. The temperature is high enough to kill growing cells and bacterial spores.

3.4 GROWING MICROORGANISMS

Microorganisms are grown, or **cultured**, in a nutrient mixture called a **culture medium**. In laboratories they are grown in petri dishes or plugged bottles or conical flasks. A small fermenter holding a few litres of culture medium can also be used (Figure 3.4). In industry large stainless steel tanks holding hundreds or thousands of litres, called **fermenters**, are used to grow microorganisms. Everything, media and containers, must be sterile before they are used.

FIG 3.4 Pilot-scale fermenters are used to investigate how well the organisms stand up to industrial conditions and to look at the yield of product under these conditions.

What microorganisms need to grow

Like all living things microorganisms need the right nutrients in the right proportions. Recipes for culture media have been developed by trial and error or from a greater understanding of a species' nutritional needs. All organisms have certain needs that any medium has to fulfil. You might find it useful to reread the sections on bacterial and fungal metabolism in Chapter 2. All microorganisms require:

- **nitrogen-containing compounds**, to provide the raw materials for protein and nucleic acid synthesis. Some use nitrogen from the atmosphere, others use inorganic ions such as nitrates, or complex organic nitrogen-containing molecules such as peptides.

- **carbon-containing compounds** for many different biochemical activities including the synthesis of new materials. Some use carbon dioxide but most use organic carbon compounds such as sugars or alcohols.

- an **energy source**. Photosynthesisers need light, chemoautotrophs need inorganic compounds to oxidise. The rest usually use carbon-containing compounds that are fed into respiration.

- There may be specific **growth factors** such as vitamins, minerals or fatty acids.

Culture media

All culture media should contain substances to match the needs outlined above. All media are based on a solution of salts. A **buffer** is added to keep the pH constant. Trace elements are usually provided by impurities in laboratory chemicals used to make the media. Only nitrogen-fixing photosynthesisers such as cyanobacteria and a few other species can grow in a solution as simple as this. When more nutrients are added, such as sugars and nitrate ions, more organisms can grow. Culture media are supplied as powders that are made up in water and sterilised before use. Once it has been used to grow microorganisms it must be sterilised by autoclaving or disinfectant before disposal.

The most widely used culture media are known as general purpose or **broad spectrum** media, which contain a selection of nutrients useful to most organisms. Examples are nutrient agar, nutrient broth and malt agar. Many species will grow on them and they are also useful for organisms whose needs are not fully understood. They contain hydrolysed meat extract, yeast extract or malt extract, which are mixtures of soluble nutrients including sugars, peptides, amino acids, vitamins, nucleic bases and a large selection of inorganic ions. Peptone, which is partly hydrolysed protein containing phosphates and sulphur as well as organic nitrogen and carbon compounds, is a common ingredient.

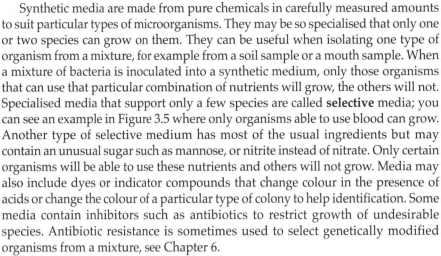

Synthetic media are made from pure chemicals in carefully measured amounts to suit particular types of microorganisms. They may be so specialised that only one or two species can grow on them. They can be useful when isolating one type of organism from a mixture, for example from a soil sample or a mouth sample. When a mixture of bacteria is inoculated into a synthetic medium, only those organisms that can use that particular combination of nutrients will grow, the others will not. Specialised media that support only a few species are called **selective** media; you can see an example in Figure 3.5 where only organisms able to use blood can grow. Another type of selective medium has most of the usual ingredients but may contain an unusual sugar such as mannose, or nitrite instead of nitrate. Only certain organisms will be able to use these nutrients and others will not grow. Media may also include dyes or indicator compounds that change colour in the presence of acids or change the colour of a particular type of colony to help identification. Some media contain inhibitors such as antibiotics to restrict growth of undesirable species. Antibiotic resistance is sometimes used to select genetically modified organisms from a mixture, see Chapter 6.

A minimal medium is one that provides only for the specific needs of a species with no other factors. Minimal media are useful in investigating biochemical pathways. Enrichment media are synthetic media supplemented with materials that promote the growth of one species more than others.

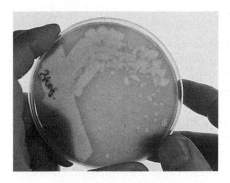

FIG 3.5 These streptococci growing on blood agar can affect red blood cells. They have broken down haemoglobin in the clear areas around the colonies.

Solid and liquid media

Organisms can be grown in liquid culture media, such as nutrient broth, or nutrient slurries for large-scale industrial processing. A small quantity, or **inoculum**, of organisms is added to the culture medium where they grow as individual cells dispersed through the solution, making it appear cloudy, or **turbid**.

For many purposes microbiologists prefer to grow microorganisms on solid or semi-solid media. A gelling agent such as agar is added to the culture medium. Agar is a gelling polysaccharide extracted from seaweed that very few organisms can metabolise so it remains solid as microorganisms grow. Agar powder is mixed with the nutrients when the medium is made up, then the medium is autoclaved. The hot mixture is poured into petri dishes or small bottles. As the medium cools to below 40 °C it sets to make a porous semi-solid surface. Solutes and extracellular enzymes are able to diffuse through it easily. When bacteria or fungal spores are spread over the surface they grow and reproduce on the surface to form clumps of cells called **colonies**, seen in Figure 3.6. Sometimes bacteria may be inoculated deep into agar dispensed in bottles. These **stab cultures** are useful for identification of unknown bacteria as some liquefy the media, change an indicator colour, or migrate out from the stab.

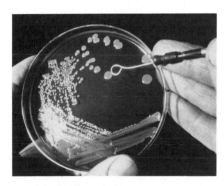

FIG 3.6 When a sample of bacteria are inoculated using a streak technique onto a nutrient agar plate, they are gradually spread out. The aim is to reach single, independent colonies, each the result of one bacterium's growth.

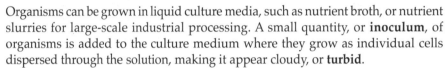

3.5 ## BATCH AND CONTINUOUS CULTURE

Typically bacteria and fungi are grown in a **batch** culture. The organisms are inoculated into a quantity of sterile culture medium in a sterile culture vessel and incubated at the optimum temperature for growth. The culture is provided with a suitable atmosphere – air for aerobic organisms, carbon dioxide and hydrogen for anaerobes, or enriched with carbon dioxide, methane, nitrogen, or hydrogen. If the medium is liquid it is stirred, shaken, or mixed by the aeration system. The microorganisms grow and multiply until the medium becomes too unfavourable or a nutrient runs low, then the culture declines and growth slows. Microorganisms, or any product they have made, can be harvested from the culture at the point that gives maximum yield for the cost incurred. After each batch is finished the process is stopped and the equipment sterilised again before the next batch. Commercial processes use batch cultures mainly, and several are described in Chapters 4 and 6.

Batch culture may not be the best way to produce cells or make a product. For example, if a steady supply of cells is needed then a young culture of the most actively dividing cells is best. Some products, for example antibiotics, are made by cells only in certain growth phases. **Continuous culture** techniques are designed to keep organisms in a particular phase continuously. The state of the culture medium within the main culture vessel is monitored constantly in order to keep the organisms growing in the desired state. If there is a shortage of a nutrient or oxygen, more is added; if the pH changes, it is adjusted by adding base or acid. The culture medium is continuously drained off to harvest the product. Industrial producers would prefer to grow microorganisms in continuous culture because there is a constant controllable supply of product and the equipment is never idle. The constant output allows manufacturers to use smaller volumes of culture. However, there may be problems maintaining optimum conditions, preventing contamination, ensuring consistency and in maintaining quality control measures. Few large-scale processes use continuous culture at present.

QUESTIONS

3.5 Write definitions of the following terms: culturing, autoclaving, inoculum, turbid, buffer, selective culture medium.

3.6 Why is it necessary to autoclave media before and after use?

3.7 Read the recipe below for a medium for growing fungi:

CZAPEK DOX MEDIUM	
Sodium nitrate	20.0 g
Potassium chloride	0.5 g
Magnesium glycerophosphate	0.5 g
Potassium sulphate	0.35 g
Sucrose	30.0 g
Agar	12.0 g
Distilled water to 1 dm^3	

Identify the nitrogen source and the carbon source. Is this likely to be a solid or liquid medium?

3.8 Distinguish between batch and continuous culture. Make a table of the advantages and disadvantages of each method.

3.6 ISOLATING BACTERIA

Microorganisms occur naturally as a mixture of species with broadly similar needs, and any sample, such as a soil solution or mouth swab, will contain a variety of species. Techniques for isolating a single species of bacterium or fungus from a mixture depend on bacteria or fungal spores forming colonies on solid media. A sample of the mixture is inoculated onto a nutrient agar plate, each microorganism grows and multiplies and forms a colony. Even motile bacteria will not travel very far in the moisture film before colony growth can be seen. As long as the initial inoculum contains a small number of organisms that are well spread out, it can be assumed that each colony is the progeny of just one bacterium. Figure

3.6 shows individual colonies. Each colony should contain a single species of bacterium, but further steps are taken to check this. Figure 3.7 shows the steps used to obtain a pure culture.

The process of taking a sample and growing on fresh medium is called **subculturing**. Some species can be identified at this stage from the type of culture medium used and the appearance of the colony, which may have very distinct colours, shapes and textures. Other species may need biochemical tests to identify them. Isolating bacteria is made easier by using a selective medium for the first culture, which reduces the number of species that grow.

FIG 3.7 Isolating a bacterium to obtain a pure culture of a single species.

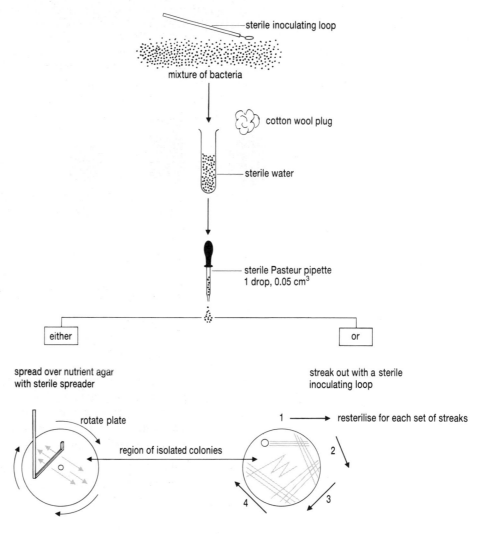

Incubate for 1 or 2 days, choose an isolated colony and take a sample. Streak out as before, incubate. There should be only one sort of colony growing. Repeat the subculture one more time.

3.7 MEASURING BACTERIAL GROWTH

Bacteria are not easy to see except with very good microscopes so it is hard to monitor their growth and numbers by sampling or direct observation. The large numbers of cells involved poses another problem, for example a sample from a bacterial colony which has been growing in nutrient broth overnight in optimum conditions can contain about 10^8 organisms per cm^3. A tiny drop from the end of a pipette with a volume of 0.05 cm^3 will contain about 5 million cells – far too many to count accurately.

Total counts

Direct counts attempt to count all the cells present in a culture. Samples of microorganisms are diluted until there are relatively small numbers of cells in the solution. These are counted, then multiplied by the dilution factor to estimate the number in the original sample. Figure 3.8 shows the process of making successive **serial dilutions** to reach countable numbers.

FIG 3.8 Making a serial dilution.

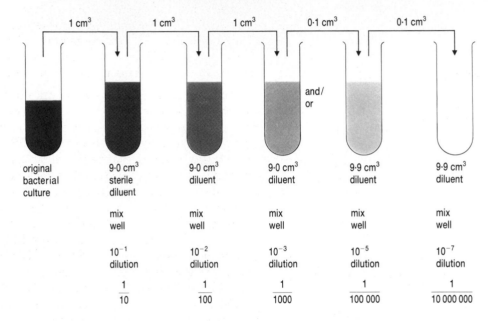

	1 cm^3	1 cm^3	0·1 cm^3	0·1 cm^3

original bacterial culture	9·0 cm^3 sterile diluent	9·0 cm^3 diluent	9·0 cm^3 diluent	9·9 cm^3 diluent	9·9 cm^3 diluent
	mix well	mix well	mix well	mix well	mix well
	10^{-1} dilution	10^{-2} dilution	10^{-3} dilution	10^{-5} dilution	10^{-7} dilution
	$\frac{1}{10}$	$\frac{1}{100}$	$\frac{1}{1000}$	$\frac{1}{100\,000}$	$\frac{1}{10\,000\,000}$

and/or

The cells in a small volume of diluted sample are counted in a suitable counting chamber. A **haemocytometer** can be used to count yeast or animal cells and a similar but smaller scale slide for bacteria. It is a thick slide with a slightly thinner central section with a grid etched on it, as shown in Figure 3.9. A coverslip is placed over the central section to form a chamber of known depth. The size of each grid square is known and so the volume of liquid over it can be calculated. Usually it is

FIG 3.9 **(a)** The haemocytometer is used to count cells. **(b)** the counting grid.

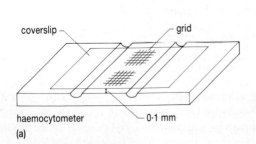

coverslip — grid

haemocytometer — 0·1 mm

(a)

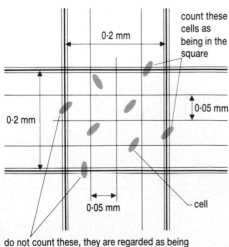

count these cells as being in the square

0·2 mm

0·05 mm

0·2 mm

0·05 mm

cell

do not count these, they are regarded as being in the squares below and to the side.

(b)

0.004 mm^3. A sample of diluted cell culture is introduced into the central section and viewed under a microscope. The number of organisms in several squares are counted, an average found and, using the dilution factor, the number per cm^3 in the original sample is estimated. Care has to be taken with cells that overlap squares, and a procedure is shown in Figure 3.9 that prevents the same cell being counted twice. Unfortunately this technique cannot distinguish between dead cells and live ones.

Example

In the diagram there are 6 bacteria in the square, which has an area of 0.04 mm^2. The chamber has a depth of 0.1 mm.
Therefore there are 6 bacteria in 0.004 mm^3.
Therefore in 1 mm^3 there are 1500 bacteria.
The sample this was taken from was diluted to a 10^{-5} dilution.
Therefore there are 1500×10^5 organisms in 1 mm^3 of original sample.
Therefore there are $1000 \times 1500 \times 10^5$ organisms in 1 cm^3 of original sample.
Expressed as standard form: 1.5×10^{11} organisms per cm^3

The effects of microbial reproduction can also be used to assess growth in **indirect counts**. As organisms grow they make a nutrient broth turbid. The turbidity is measured using a **colorimeter**. This passes a beam of light through a culture sample and the light absorbed or scattered is recorded. The more organisms there are the more light is absorbed and the greater the optical density of the solution. Another method uses a Coulter counter, originally devised for blood cells. A probe with two electrodes is put in a sample of culture. One of the electrodes is inside a small glass tube with a narrow entrance. When a bacterium passes through the entrance it alters the conductivity inside the probe and is recorded. The size of the alteration depends on the size of the bacterium so a cell size analysis can be done. Counts can also be done using a laser beam; as cells in a suspension pass through, they interrupt the beam of light and are recorded.

These methods cannot distinguish live cells from dead ones, nor can they distinguish cells from particulate matter with a similar size, but they can be automated for the rapid analysis of a large number of samples.

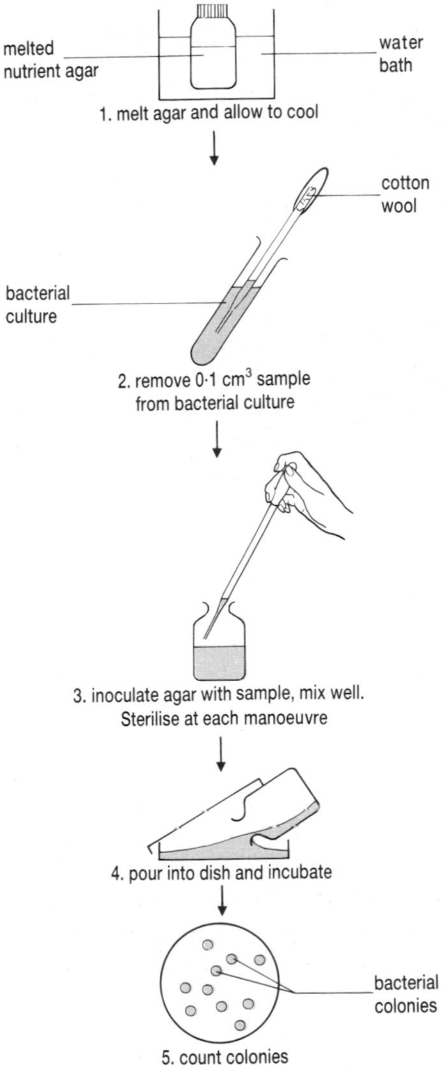

Viable counts

Only cells capable of growing are counted in **viable counts**. A **plate count** is used to count bacteria; the procedure is shown in Figure 3.10. Samples are serially diluted before counting. Several suitable dilutions are made and, at each dilution, a measured volume of culture is mixed with melted but cool nutrient agar and poured into a petri dish. Each living bacterium in the suspension will form a colony. Duplicate plates are prepared for each dilution to give a more accurate estimate. The dishes are incubated until colonies appear, then the number of colonies is counted. This is multiplied by the dilution factor to give an estimate of the number of bacteria in the original sample.

melted nutrient agar — water bath
1. melt agar and allow to cool

cotton wool
bacterial culture
2. remove 0·1 cm^3 sample from bacterial culture

3. inoculate agar with sample, mix well. Sterilise at each manoeuvre

4. pour into dish and incubate

bacterial colonies
5. count colonies

FIG 3.10 The pour-plate method to count viable bacteria.

QUESTIONS

3.9 Distinguish between a total count and a viable count.

3.10 If there were 7×10^5 organisms per cm^3 in a suspension, how many organisms would you expect in 1 cm^3 of a 10^{-3} dilution?

3.11 The following counts were obtained from a series of plate counts, each using a 0.1 cm^3 sample per plate, to estimate how many bacteria were in a water sample. At a dilution of 10^{-5} there were 87, 73, 91 colonies; at 10^{-6}: 9, 8, 8; and at 10^{-7}: 1, 0, 0. Work out the average count at **each** dilution. Use the figure for each dilution to calculate how many bacteria were present originally. Can you suggest a reason why they are not the same? Use your figures to estimate how many viable bacteria were in the original culture.

3.8 # MEASURING FUNGAL GROWTH

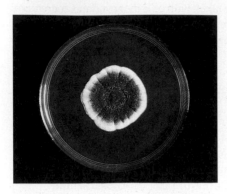

FIG 3.11 One way to measure fungal growth is to monitor the increase in mycelium diameter.

Yeasts are unicellular so the techniques for counting bacteria are useful. Monitoring the growth of mycelial fungi is different. The diameter of the mycelium can be used to monitor growth in a petri dish. A small sample of mycelium will grow by producing hyphae that branch repeatedly, radiating out in all directions. If there is an even distribution of nutrients, for example on an agar plate, a circular mycelium develops, like that in Figure 3.11. Hyphae grow from their tips so the diameter of the mycelial mat increases as the fungus grows.

Fungi are grown in liquids for large-scale industrial production, like those used for growing yeast and bacteria. Fungi in liquids normally grow as floating masses of hyphae at the air–liquid interface. However, maximum growth rate is reached by growing fungi in a **submerged culture**. The vigorous aeration for high oxygen concentration also mixes the culture medium and breaks up the fungal mycelium into small pellets. Each pellet has a high surface area to volume ratio that allows efficient gas and nutrient exchange, and is capable of exponential growth. Growth is measured by the change in fungal dry mass over a period of time. Samples of the culture are taken at intervals, the mycelium is filtered out, dried and weighed, and growth is measured as the increase in grams per gram dry mass in unit time. Bacteria grown in liquid cultures can be monitored in this way too.

3.9 # THE BACTERIAL GROWTH CURVE

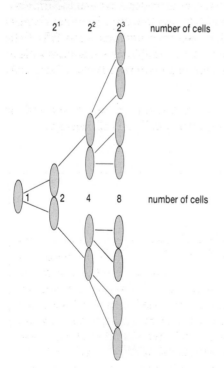

2^1 2^2 2^3 number of cells

1 2 4 8 number of cells

FIG 3.12 When bacteria multiply in favourable conditions, the numbers grow exponentially.

Microbial growth usually refers to changes in the size of a population, not a change in the size or mass of an individual. Bacteria reproduce by binary fission. Each cell divides into two smaller cells, each of which goes on to divide into two more cells and so on, as shown in Figure 3.12. The numbers of cells double each generation. If the environment provides plenty of nutrients and other materials needed for growth, bacteria grow very quickly and binary fission is rapid. The bacteria divide before they reach maximum size and the life span of a cell, called its **generation time**, decreases. The cells are in their most active metabolic state. Bacteria multiplying at their maximum rate are described as being in a state of **balanced growth**. All measures of cell growth, such as increase in dry weight, DNA content and protein content, double each generation. The population descended from *each* bacterium after a number of generations of balanced growth is estimated using the general equation:

population in the nth generation $= 2^n$

Within a few hours such large numbers are generated that they are plotted as logarithms on a graph. This exponential growth can continue indefinitely – but only if the environment is favourable. In natural environments a shortage of a nutrient, accumulation of toxic materials, a change in an environmental factor or predators slows population growth.

The growth curve

A growth curve shows how a population of bacteria in a batch culture changes with time. A batch of sterile nutrient broth is inoculated with bacteria and incubated at optimum temperature. Initially the organisms have plenty of nutrients, salts, oxygen or other materials needed for respiration, and growth factors. Samples are

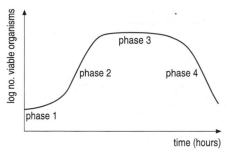

FIG 3.13 The bacterial growth curve.

taken at regular intervals and the numbers of viable bacteria present are counted. The numbers of bacteria are plotted as log numbers against time, giving a growth curve like the one shown in Figure 3.13.

The bacteria use materials in the culture medium as they grow, and secrete and excrete metabolites into it. Consequently the medium's composition is constantly changing and this affects bacterial growth. The growth curve consists of different phases that reflect changes in bacterial metabolic activities through time. The phases change gradually from one to another.

Phase 1 Lag, latent or initial stationary phase. The bacteria are active but reproduce very slowly so numbers do not increase much. They imbibe water, synthesise ribosomes and induce enzymes needed to exploit the culture medium. The length of this phase depends on the medium the bacteria grew in before the investigation started, and how they were growing. It is very short if the cells were actively growing in a similar medium, but long if the bacteria were taken from a stock culture in storage. At first, large cells with long generation times are produced but gradually the generation time gets shorter and the cells smaller as they enter the next phase.

Phase 2 Exponential or log phase. In this phase bacteria reproduce at their fastest and numbers increase directly with time. The generation time is very short, though it varies with species. *E. coli* doubles its numbers every 21 minutes at its fastest, and *B. stearothermophilus* every 9 minutes. The number of doublings per unit time (usually an hour) is the **exponential growth rate constant**. The bacteria have a very active metabolism in this phase and are in balanced growth. It is the most useful phase for investigating growth.

Phase 3 Stationary phase. As the cells grow they alter the culture medium. Nutrients and materials needed for respiration become depleted, and there is a fall in pH as carbon dioxide, acids and other metabolites build up. There are changes in the cells as their energy stores are used up, and the reproductive rate falls. Cells may begin to produce secondary metabolites such as antibiotics. Organisms die in greater numbers so there is no overall increase in growth rate. The surviving cells are better able to survive in the difficult conditions.

Phase 4 Death phase. Conditions in the culture medium are very severe and far more bacteria die than are produced, so the number of living cells declines.

 SPOTLIGHT

Diauxic growth

Outside the laboratory microorganisms usually have several energy sources available to them. First they use one for which they have the necessary enzymes already and which requires the least metabolic effort to use, growing exponentially until its concentration runs very low. The presence of another energy source causes the induction of enzymes needed to metabolise that source during a second lag phase. Growth will then restart on a second exponential phase; this is **diauxic** growth. Jacob and Monod did their Nobel prize winning work on gene regulation on the lac operon. They used *E. coli* cells actively growing on glucose and transferred them to a lactose-containing medium. The bacteria stopped reproducing because they did not have the necessary enzymes to use lactose. The genes that encoded the three enzymes necessary to use lactose were repressed by a protein made by a regulator gene. The presence of lactose affected the repressor protein so that it could no longer repress the lactose metabolising enzymes.

QUESTIONS

3.12 You are provided with a culture of bacteria. Outline a method you could use to determine the number of living cells per cm³ in the culture.

3.13 The graph in Figure 3.14 shows the growth of a culture you are investigating. Log numbers of living cells have been plotted against time. The culture medium is a minimal medium with a single carbon source.

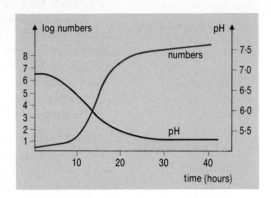

FIG 3.14

(a) Explain the changes in the curve of numbers of living cells between 5 hours and 15 hours, and between 25 and 40 hours.

(b) The pH of the culture medium has been plotted. Suggest one substance that could have accumulated during the course of the investigation and brought about the change in pH?

(c) What would happen if another carbon source were added after 36 hours?

(d) Draw a graph of the results you would have expected if the total number of organisms had been recorded instead of the numbers of living organisms.

3.14 Describe how you would set up a culture of bacteria in a petri dish starting from a stock culture. Include all the safety precautions you would take.

3.10 GROWING AND COUNTING VIRUSES

Viruses cannot grow on nutrient agar because they only replicate inside cells. Viruses are grown in a culture of animal cells called a **tissue culture**, which is a flat sheet of host cells on the bottom of a petri dish. You might find it useful to reread Section 2.6 on viruses.

Tissue culture

Many kinds of cells and tissues can be taken from living organisms and persuaded to grow in the laboratory. They will grow in a specialised cell culture medium if kept at the correct temperature and pH.

A suspension of cells is used to make a tissue culture. They are mixed with a suitable sterile cell culture medium and dispensed into sterile petri dishes or bottles where they settle and divide. New cells move around but when they contact another cell they stop, so a layer just one cell thick called a **monolayer** forms within a few days. The culture medium is a complex solution of salts with glucose and vitamins. Other factors are needed in small amounts and are supplied by adding about 10% horse or calf serum. Animal cells grow within a very narrow temperature range and need an environment with the same concentration of CO_2 as tissue fluid, higher than in the atmosphere. An indicator in the medium gives a quick visual check of pH.

Growing and counting viruses

A sample of virus-containing suspension is mixed with a suspension of suitable host cells in tissue culture medium. Viruses will infect cells in the suspension. The mixture is dispensed into tissue culture dishes or bottles and the culture left to grow. After a few days large numbers of viruses can be extracted from the culture. There will be free virus particles in the medium, and more obtained by breaking open the cells.

A plaque assay is used to estimate virus numbers. Once a virus has multiplied inside a cell its progeny escape, damaging the cell as they go. Neighbouring cells are infected and damaged in turn as the infection spreads further away from its initial focus. Cells are so minute that at first the damage is hard to see, but after a day or so the holes in the cell layer are big enough to be seen when it is stained. Each hole is called a **plaque,** and it is assumed that each plaque is the result of one virus causing the initial infection. Plaques can be seen in Figure 3.15.

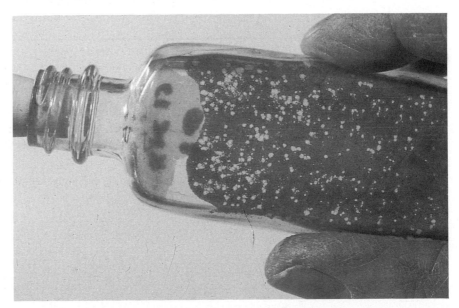

FIG 3.15 A culture of animal cells infected with a virus. The viruses destroy the cells they infect leaving plaques, or holes, in the cell sheet, which can be seen when the cells are stained.

Viruses are counted by making serial dilutions of the virus suspension, and measured volumes of each dilution are mixed with a suspension of suitable host cells in tissue culture fluid. The mixture is dispensed into a petri dish and incubated for 3 days. The sheet of cells is stained and the number of plaques at each dilution counted. The number in the original suspension can then be estimated. Bacteriophages can be counted in a similar way. They are mixed with a bacterial suspension that is spread evenly over the surface of nutrient agar in a petri dish. The bacteria grow over the agar surface in a dense 'lawn' except where the virus release has broken them open and left gaps. Each gap is assumed to be the result of one bacteriophage and its progeny.

The virus growth curve

Viruses do not grow once they have been assembled. After a virus leaves a host cell it adsorbs onto another host cell nearby and infects. Once it has lost its protein coat, a virus cannot be detected by a plaque assay because it needs the protein coat to attach to cells in a monolayer. Inside the cell, virus replication quickly generates large numbers of new infectious virions. When these escape there is a sudden rise in the numbers of virus detected. These can reinfect and repeat the cycle. When virus numbers are measured they increase in a number of steps corresponding to reproductive cycles. Figure 3.16 shows a typical growth pattern for a bacteriophage, called a **one-step growth curve**.

FIG 3.16 A bacteriophage one-step growth curve.

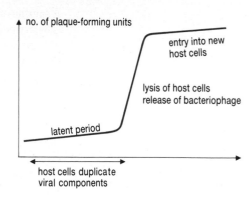

QUESTIONS

3.15 How does an indicator monitoring pH inform us about dissolved CO_2 concentrations in the tissue culture medium?

3.16 How would you use bacterial culture techniques to investigate the activity of a newly discovered antibiotic against a bacterium, *Bacillus megaspottium*, isolated from a local patient? How would you modify your investigation if you were looking at an anti-viral agent? How could you investigate whether the antibiotic had toxic effects on human cells?

3.17 The graph in Figure 3.17 shows the growth of a sample of fungus found growing in an old unwashed coffee cup. It was cultured on two different media at 25 °C for 7 days.

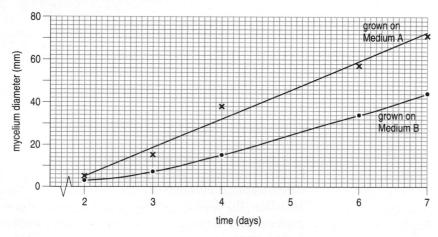

FIG 3.17

(a) On which medium did the fungus grow best?

(b) Explain why different growth patterns were obtained.

(c) How would you obtain a pure culture of this fungus?

Exam questions

3.18 The yeast *Candida utilis* was grown in a liquid culture medium. Every hour, a 1 cm³ sample was taken from the culture and diluted 100 times. One drop of the resulting suspension was then placed on a haemocytometer slide. The graph shows the mean number of yeast cells in each 0.004 mm³ of the haemocytometer grid at different times after the yeast was added to the medium.

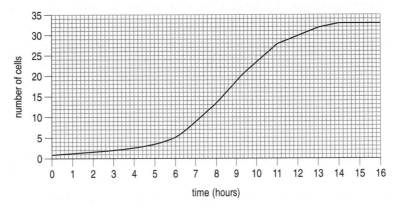

(a) Explain **one** reason for each of the following: (i) the slow initial rise in the number of yeast cells; (ii) the rapid rise in numbers that followed. (2)

(b) Use the graph to calculate the number of yeast cells per cm³ in the yeast culture at 13 hours. Show your working. (2)

(c) Describe how you could find the number of live yeast cells in 1 cm³ of medium at 13 hours. (3)

AQA Biology, March 1999, Paper BY06, Q. 4

3.19 Bacteria were grown in liquid culture. The growth of bacteria was monitored using a colorimeter, which measures the amount of light passing through the culture.

(a) Explain how, by using a colorimeter, it is possible to follow the growth of a bacterial culture over a period of time. (3)

(b) Describe and explain **one** method that you might use to find the numbers of live bacterial cells in the culture at one particular time. (3)

AQA Biology, June 1999, Paper BY06, Q. 4

3.20 The antibiotic penicillin is produced by a species of fungus which is grown in a fermenter. The table shows the change in concentration of the fungal biomass, carbohydrate and ammonia during the course of fermentation.

TIME/hours	CONCENTRATION/g dm⁻³		
	FUNGAL BIOMASS	CARBOHYDRATE	AMMONIA
0	8	100	33
10	8	88	29
20	21	70	20
30	24	35	13
40	27	21	9
50	30	4	6
60	32	1	4
70	33	0	3

(a) Calculate the mean growth rate of the fungus between 30 hours and 50 hours. Show your working. (2)

(b) **(i)** Explain why the growth rate was slower between 0 and 10 hours than it was between 10 and 20 hours. (1)

(ii) Use the data in the table to explain why the growth rate of the fungus was very slow after 60 hours. (2)

AEB Biology, January 1999, Paper AS-A/MBIO/2, Q. 2

SUMMARY

Microrganisms are widely distributed in all environments. They are easily dispersed and rapidly exploit a favourable niche.

Their growth is affected by pH, water potential, oxygen and substrate concentrations. Individual species grow best within a particular range of temperatures and are adversely affected outside these. Growth is rapid when all factors are present at optimal values.

Pure cultures of microorganisms contain only one sort of organism. Microorganisms are selected from mixtures using techniques to spread out the cells on media that favour the growth of one species over the others.

All items used to grow microorganisms must be free of contaminating organisms. Culturing microorganisms requires aseptic techniques to keep out contaminants.

Microorganisms can be grown in artificial culture media and the precise composition determines which organisms can grow on it. Broad spectrum media contain a wide variety of nutrients in order to grow a wide variety of organisms.

Microorganisms and cells can be grown in batches, or they can be maintained in a particular growth phase in continuous culture.

Microorganisms can be counted directly or their numbers estimated by their effects. Some methods distinguish between living and non-living cells. Viable counts count organisms that are capable of growing; direct counts count all the organisms present in a sample of culture.

Bacteria and viruses have characteristic growth patterns. When cells are in a state of balanced growth they multiply exponentially.

ENVIRONMENTAL AND INDUSTRIAL BIOTECHNOLOGY

"Mr Bosher, upon being appealed to for his opinion, explained that science was alright in its way, but unreliable: the things scientists said yesterday they contradicted today, and what they said today they would probably repudiate tomorrow."

Robert Tressell, The Ragged Trousered Philanthropists

Mr Bosher, speaking 100 years ago, was right, but not in the way he intended. Scientific advances are made daily. Mr Bosher would have been astounded to find out that today a whole forest can be raised in an incubator the size of a domestic fridge-freezer, or that bacteria regularly make human hormones.

New advances have profound significance. Developments such as the large-scale manufacture of antibiotics by fungi have transformed our lives. Industries using microorganisms have burgeoned and many exciting new products have been developed. Our ability to manipulate a cell's genetic material allows us to do 'impossible' things. This section is about using microorganisms to make products and deal with wastes.

PREREQUISITES

Refresh your understanding of Sections 2.2, 2.5–2.6 and Sections 3.2–3.5 and 3.7–3.9, which cover the biology of commercially grown organisms and how they are grown. An understanding of GCSE chemistry and enzyme biology is helpful, but no previous knowledge of food science is required.

4 FOOD MICROBIOLOGY

LEARNING OBJECTIVES

After studying this chapter you should be able to:

① explain why raw foods have microorganisms in them

② describe how yeast and lactic acid bacteria are used in food and drink production

③ describe the use of other microorganisms and their products in food production

④ explain how microorganisms spoil food

⑤ understand how microorganisms in foods can cause ill health

⑥ describe the main ways of preserving food.

4.1 MICROORGANISMS AND FOOD

Microorganisms in our natural environment find their way onto the food we eat. Many bacteria and fungi have similar nutritional needs to our own, so our food is suitable for their growth too.

Microorganisms alter food by secreting enzymes into the food they are growing on and absorbing the small soluble nutrient molecules produced by the breakdown of large complex organic molecules. This enzyme action changes the food's appearance and texture. Proteins, fats and carbohydrates are broken down into smaller molecules, altering the flavour, the smell and the texture. There might be obvious bacterial colonies or fungal mycelia. Surprisingly, sometimes we want these changes, for example for making milk into cheese (Section 4.5). At other times, though, the changes are undesirable; food may become slimy, develop 'off' flavours and unpleasant smells, and the texture breaks down. Food production, processing and storage is carried out using carefully controlled strains of microorganisms in a way that encourages desirable changes and reduces the growth of contaminating organisms. Certain microorganisms growing in food cause disease by producing toxins, or cause food poisoning (Sections 4.9 and 8.4).

4.2 WHERE MICROORGANISMS IN FOOD COME FROM

FIG 4.1 Wild yeasts on the grapes make a greyish film or 'bloom'. At one time the bloom was the source of yeast for wine making.

Microorganisms in foods come from a number of sources. The outer surface of fresh fruit and vegetables carry bacteria, fungi and yeasts (seen in Figure 4.1) from the fields and the soil in which they were grown. As long as the outer layers remain intact, these organisms grow very little and do not present problems. Similarly the outer surface of fresh meat, chicken and fish carry organisms from the animal's environment, or from the gut. These generally stay on the surface and do not penetrate into the food.

SPOTLIGHT

Organisms on food

A tomato taken straight from the plant may have millions of organisms on each cm^2 of surface. This is reduced to a few hundred after washing. A cabbage straight out of a field has between 1 and 2 million organisms per gram of outer leaf but after washing and trimming this number falls to less than a quarter. There are only a few hundred per gram right in the centre of the cabbage.

Wherever food is exposed to the air – on market stalls, shop counters or kitchen worktops – it gathers airborne organisms, particularly fungal spores. More organisms arrive with flies that land on the food. Handling and processing food adds more unwanted organisms; the more processing, the more likely the contamination from people's clothes and skin. **Cross-contamination** occurs when microorganisms are transferred from one food to another, for example when fresh meat is stored next to cooked meats.

QUESTIONS

4.1 Where do microorganisms found in and on food come from?

4.2 Define cross-contamination.

4.3 List three ways in which microorganisms can change food.

4.4 Describe how microbial enzymes can change food.

4.5 State one way in which microorganisms on food can be harmful to human health.

4.6 This question is quite hard. You will need to think carefully and put together what you have just learnt and information from your general knowledge.

Do you think there are many bacteria in each of the following foods? (a) an orange, (b) a carton of pasteurised milk, (c) a vacuum packed kipper, (d) roast beef carved then left covered overnight, (e) a carton of natural yoghurt.

When you have read **all** of this chapter, look at your answers again. Alter your answers if necessary.

4.3 MAKING FOODS WITH MICROORGANISMS

World wide there are traditions of making foods using microorganisms. Bread, beer, wines and spirits, olives, sauerkraut, yoghurt, cheese, fermented milk drinks, buttermilk, vinegar, soy sauces, coffee, tea and cocoa are all processed using microorganisms. The techniques prolong the life of perishable foods so they can be stored against times of scarcity. Milk can be kept for longer when made into butter, yoghurt or cheese. Sauerkraut, pickles and silage (used for animal food) were used in late winter when green vegetables were unobtainable. Foods processed by microbes develop different flavours too: fermented soy sauce tastes quite different to soy beans, as well as having a different nutrient content.

Though the people who developed these foods had very little idea of the biological processes involved, the traditional methods manipulated environmental conditions to bring about the preferential growth of certain organisms. Many modern techniques are large-scale refinements of the traditional methods, with the emphasis on factory hygiene, maintaining pure stable cultures of microorganisms (Chapter 3 explains pure cultures) and precise control of manufacturing conditions.

Microbial products are also used to make ingredients used in food and drinks manufacture, such as citric acid used in soft drinks and sweets, glucose syrups for bakery products, and thickening agents for low calorie foods. Mycoprotein, sold as Quorn, a protein food, is made from microorganisms growing on wastes from other food industries. Check your specifications for the processes you need to study in detail.

4.4 YEAST AND ITS PRODUCTS

Using yeast to make foods and drinks is probably the oldest form of biotechnology. Bread and fermented drinks have been with us for thousands of years. Modern methods use varieties of the yeast *Saccharomyces*. Yeast carries out aerobic respiration when there is plenty of oxygen available and makes large quantities of carbon dioxide; in low-oxygen environments it produces less gas and more ethanol from anaerobic respiration. You can find out more about the biology of yeast cells in Chapter 2.

Bread

Enzymes survive the milling process when grains are made into flour. These enzymes act on the starch in bread dough to make a mixture of sugars including maltose and glucose. Bread flours often also contain a fungal amylase to enhance the breakdown of starch to glucose. Yeast uses the sugars in aerobic respiration, which generates carbon dioxide, making the dough rise as bubbles of gas are caught in the elastic dough. Water and ethanol produced at the same time are driven off during baking. In the Chorleywood process Vitamin C is also added during dough making to aid yeast growth and reduce the amount of time the bread proves. Lactobacilli may also grow during the early stages of proving, generating lactic acid; this contributes to the final flavour and inhibits other organisms. Bakeries may add other ingredients such as whiteners, raising agents, stabilisers and flavourings.

Beer

Beers differ in the types and proportions of ingredients used and the treatment each ingredient undergoes. The brewing process is summarised in Figure 4.2; essentially starchy materials are converted to sugars that are then fermented to ethanol and carbon dioxide.

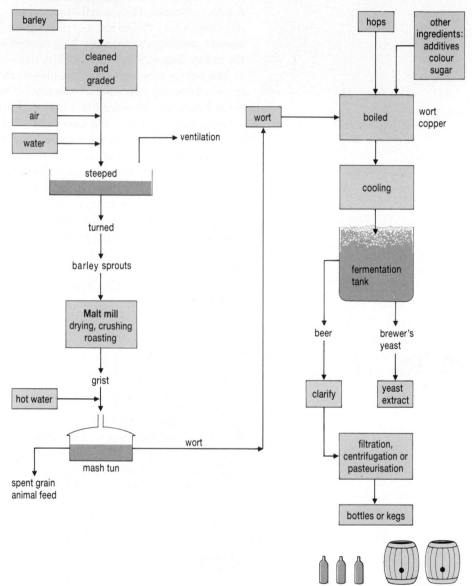

FIG 4.2 The brewing process.

Grains of barley, rice or maize are moistened and allowed to sprout. During germination enzymes in the grains convert the starch reserves into smaller molecules. The sprouted grains are dried, crushed and roasted to give a dark colour and rich flavours to the beer. More enzyme action in the mash tun produces monosaccharides and simple nitrogenous compounds. The nutrient-rich liquor is called **wort**. The higher the sugar content of the mash the more alcohol is eventually made. Some brewers add enzymes such as amyloglucosidase and α-amylase to the mash to get as much sugar as possible. Hops, or hop extract, and colourings are added before the wort is boiled. Hops give flavour; they also have anti-microbial activity, which reduces the growth of other microorganisms that could spoil the beer. Hygienic conditions must be carefully maintained to stop spoilage bacteria entering and growing in the beer.

After cooling, the mix is run into a deep fermentation tank (see Figure 4.3) and inoculated with a cultivated strain of *Saccharomyces*, often saved from the previous batch of beer. There are a multitude of strains of yeast used in brewing, mostly *S. cerevisiae* and *S. carlsbergensis*, but there are others. New yeast starter cultures are made up from laboratory stocks periodically to ensure consistency between brews. The deep tank and the layer of carbon dioxide generated by yeast activity results in a low oxygen concentration in the tank, so anaerobic fermentation is ensured, producing an ethanol concentration of up to 6%. After the fermentation is stopped, beer is separated from the yeast by centrifuging. The beer is clarified, treated to prevent spoilage and packaged.

Yeast and carbon dioxide are valuable products in their own right. Yeast is a source of vitamins and can be made into yeast extract for food processing and savoury spreads. Carbon dioxide is bottled under pressure for a variety of uses.

Wine making is a similar process. Wines made with local yeast strains give particular flavours and aromas to the wine, different to wine produced elsewhere. Both table wines and wines for distilleries use specific scientifically cultured strains of yeast, to give consistency. Figure 4.4 outlines wine production.

FIG 4.3 Beer is brewed in deep tanks with the minimum surface exposed to the air, to reduce oxygen concentrations in the beer. Fermenting yeasts quickly generate a blanket of carbon dioxide just above the beer which excludes air.

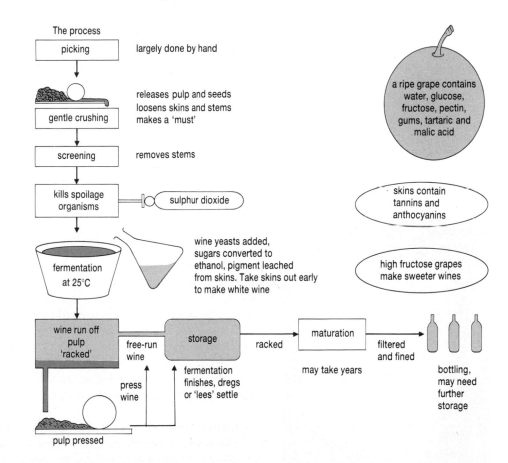

FIG 4.4 How wine is made.

FOODS MADE WITH LACTIC ACID BACTERIA

Lactic acid bacteria are used in many food processes. They are fermenting organisms that do not need oxygen but do need high nutrient concentrations, which limits them to particular nutrient-rich environments. They grow well in milk, foods rich in sugars, vegetation, intestinal tracts and on mucous membranes. Though lactic acid bacteria may already be present in the raw food, they are usually added as pure cultures during processing.

Sauerkraut

Naturally occurring lactobacilli predominate in sauerkraut production. Cabbage is shredded, packed with salt and pressed to exclude air. Carbon-rich and nitrogen-rich nutrients leach from the leaves, and the virtually anaerobic environment is ideal for lactic acid bacteria. There are lactic acid bacteria in the normal leaf flora but a starter culture may be used. *Leuconostoc* grows first but is superseded by *Lactobacillus plantarum* over a period of weeks. Bacterial activity quickly generates acidity of pH 5 or less, which inhibits other organisms. Silage is grass that has been preserved in a similar way, for feeding livestock. An acid environment stimulates the growth of lactic acid bacteria and reduces the growth of other organisms.

Dairy products

In cheese and yoghurt manufacture, acids made by the bacteria lower the milk's pH, which in turn coagulates milk proteins. Lactic acid bacteria such as lactobacilli and streptococci are naturally found in a cow's milk ducts, but milk is usually pasteurised (see Section 4.10) before being used for cheese or yoghurt. Pasteurising kills most wild bacteria that could spoil the milk or cause human illness. Scientifically raised starter cultures of lactic acid bacteria are added to pasteurised milk at the start of production.

Cheese

Lactobacillus and *Streptococcus* species are inoculated into warm milk and left to incubate. If another microorganism is necessary for ripening it is added too, for example the fungal spores needed for Stilton cheese. Rennet or chymosin is also added to improve curdling. During incubation the bacteria ferment lactose, the sugar in milk, to lactic acid. This lowers the pH and affects the structure of milk proteins. Proteins coagulate into curds, which also trap fats in the milk. After incubation, the jelly-like curds are cut up and the remaining liquid, called whey, is drained out. Whey contains milk sugars and other useful nutrients so it is dried and used in other food products, converted into a sweet syrup, or used as animal feed.

Manipulating the environmental conditions, together with careful use of heat and salt, converts curds into the different kinds of cheese. After salt has been added the curds are pressed and the cheese is left to ripen. Ripening involves bacteria or fungi that break down proteins and fats. The by-products, which include butyric, propionic, succinic and diacetyl acids, esters and amines, give the cheese its flavour and aroma.

Cheeses with 'eyes', for example Gruyere, come from the action of *Propionobacterium* producing carbon dioxide that is trapped as bubbles. Blue cheeses have a culture of fungal spores added at the start of manufacture but they do not grow well in the physical conditions inside the cheese. Fungi need high oxygen concentrations so cheeses are pierced after a storage period to allow air into the centre of the cheese, as shown in Figure 4.5. This starts fungal growth and ripens the cheese for market. Cheeses such as Brie and Camembert have a rind formed by organisms wiped onto the outside. The organisms secrete enzymes that penetrate the cheese and produce flavour chemicals. Lipases extracted from microorganisms in industrial cultures can be used to shorten ripening time.

FIG 4.5 There is a heavy demand for blue cheeses at Christmas. Cheese makers can choose when to stimulate fungal growth so that the cheese will be in peak condition for the market.

Chymosin

Rennet is a mixture of two proteases that act on casein molecules in milk to coagulate them. Rennet is mostly chymosin with a small proportion of pepsin. A worldwide shortage of good quality rennet led to the development of genetically modified microorganisms carrying an artifical gene for chymosin. These make microbial rennet, now widely used in cheese making. There are also fungal proteases for making cheeses, though these degrade protein more than microbial chymosin while the cheese matures and can lead to different flavours.

If you are not familiar with the techniques for making and transferring genes, you may find it helpful to read Section 6.5. The gene for chymosin is an artificial gene, made of cDNA. The start material was mRNA for a chymosin precursor molecule, extracted from gut cells that make it. Reverse transcriptase was used to make a cDNA version of the encoded information. The artificial gene was then transferred into microbial cells using a plasmid. The gene was transferred into a number of 'easy to grow' microorganisms, but *Kluyveromyces lactis*, a yeast, is a major source. During extraction and processing the precursor molecule synthesised by the yeast is converted into active chymosin. Chymosin synthesised by these cells works identically to calf chymosin and is used to make a variety of cheeses without the use of any animal product. Microbial chymosin, like other enzymes used to process food, had to undergo a stringent series of safety tests before it was licensed for use.

Yoghurt

The biochemistry of yoghurt production is similar to that of cheese. Suitable lactic acid bacteria are inoculated into milk and the lactic acid they produce coagulates milk proteins and thickens the yoghurt.

Milk is processed to improve the yoghurt's final texture before it is inoculated with bacteria. Concentrated milk or milk powder is added to homogenised milk to increase and standardise the protein and carbohydrate content. Milk is also heat treated to denature milk proteins, which gives a more stable yoghurt as well as eliminating bacterial contaminants and reducing the oxygen content of the milk. Stabilisers to stop the yoghurt separating and sweeteners may also be added. After cooling, *Streptococcus thermophilis* and *Lactobacillus bulgaricus* are added to the milk in a carefully balanced starter culture. There should be slightly more streptococci than lactobacilli, otherwise the yoghurt becomes too acid. The bacteria ferment milk differently, acting in a mutualistic relationship, each releasing nutrients the other needs as they metabolise the milk. *S. thermophilis* needs peptides made by *L. bulgaricus* from casein in milk to grow; in turn, methanoic acid made by *S. thermophilis* encourages lactobacillus growth. *L. bulgaricus* does most of the conversion of lactose to lactic acid and makes many of the flavour and aroma chemicals. As the yoghurt culture incubates at 43 °C the pH drops. When it reaches the correct acidity the yoghurt is cooled to reduce bacterial growth. 'Bio' yoghurts use *L. acidophilus* and *Bifidobacterium bifidum* incubated at a lower temperature for longer, and have a milder taste. Once fermentation has finished the yoghurt is stored at 4 °C. Fruit yoghurts have a higher sugar content and less acidity so they have a shorter storage time. The process can be seen in Figure 4.6.

FIG 4.6 How yogurt is made.

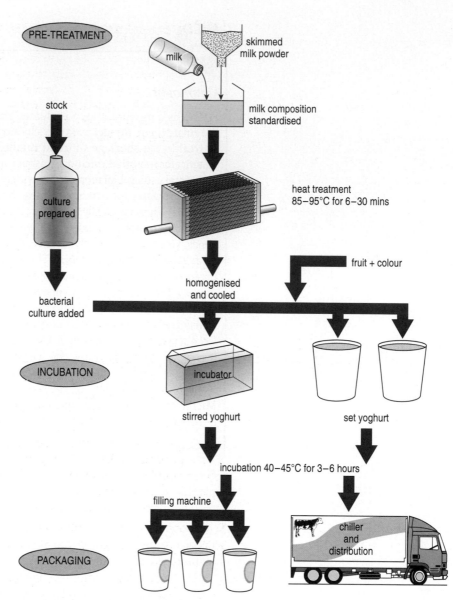

PRE-TREATMENT

skimmed milk powder

milk

stock

milk composition standardised

culture prepared

heat treatment 85–95°C for 6–30 mins

fruit + colour

homogenised and cooled

bacterial culture added

INCUBATION

incubator

stirred yoghurt

set yoghurt

incubation 40–45°C for 3–6 hours

filling machine

chiller and distribution

PACKAGING

4.6 *ASPERGILLUS* FERMENTATIONS OF SOY BEANS

Processed soy beans are a protein-rich staple food in many parts of the world. They are versatile as they can be prepared in many ways to make foods; Table 4.1 lists some of them. Soy sauce is used to flavour and colour foods.

TABLE 4.1 Microbially processed soy foods

FOOD	SOURCE	WHERE USED
miso paste	cereals and soy beans using *Aspergillus oryzae* and *Saccharomyces rouxii*	Japan, China, other Asian countries
natto	soy beans using *Bacillus subtilis*	Japan
shoyu (soy sauce)	soy beans and wheat using *A. oryzae*, *Lactobacillus*, *Hansenula* and *Saccharomyces*	China, Japan
sufu	soy beans using *Actinomucor elegans* and *Mucor* spp	China, Taiwan
tempeh	soy beans using *Rhizopus oligosporus*	Indonesia

Adapted from Rackis (1979) *Tropical Foods: Chemistry and Nutrition*, Vol. 2, ed. Inglett and Charalambous.

Industrial production of soy sauce

The type of sauce varies with the ingredients and microorganisms used but all involve a process in which bean and cereal proteins are broken down to amino acids and small peptides by microbial enzymes. The fungi *Aspergillus oryzae* and *Rhizopus*, together with lactobacilli and yeasts, are the main organisms used.

Soy beans are soaked and boiled, and wheat is roasted and crushed. They are mixed together and spread out to provide good aeration. The mixture is inoculated with a fungus known as 'Koji', usually a culture of *A. oryzae*. The fungus grows for 2–3 days on the mixture, then it is mixed with salt water to make a mash. The salt content is regulated and the mash is left in deep cool fermentation tanks for several months.

The conditions in the fermentation tanks encourage the growth of lactobacilli and yeasts. Vigorous lactic acid and alcohol fermentations take place and the mixture is aged to make raw soy sauce. This is filtered and clarified and the sediment is pressed to use as animal feed. The filtrate is pasteurised, which gives more colour and a stronger flavour, and the sediments and oils are removed. This refined sauce can then be bottled.

4.7 SINGLE CELL PROTEIN

The world shortage of protein foods has been a problem for years. The main protein supplement is soya meal, which most countries have to import and use to supplement animal foods. However, using grain and protein to feed animals to make protein is wasteful; it makes more sense to use protein for people directly. Microorganisms can have a high protein content and the idea of using them as a protein supplement has been around for a long time. Algae and *Spirulina* are gathered for food in some parts of the world but the habit isn't widespread. There were a number of problems to solve:

- the organism has to be safe to eat,
- the substrate material it grows on must be cheap and preferably not a regular food source,
- the product must be easily processed into an acceptable food that does not have untoward effects on the consumer,
- there must be no contaminants from the substrate passing through into the food,
- there must be no unusual constituents that might cause long-term damage,
- people who have been eating mushrooms, soy products, yoghurt and cheese all their lives have to be persuaded to eat microbial proteins.

Protein-rich food made by microorganisms is described as single cell protein (SCP). There are several processes using different microorganisms growing on different substrates to make SCP for animal feed. In the EU any company that wishes to introduce a new food made from materials not usually used for human consumption, or made by new processing techniques, has to obtain a pre-market clearance from the appropriate regulatory body.

Fungal SCP

Many of the SCP processes use wastes that could not otherwise be used as food and convert them into food supplements for animals. These processes can also solve local pollution problems. There are a number of processes, like one developed in Finland which uses carbohydrate-rich wastes from wood and paper industries. This waste, normally a water pollutant, is used to grow a fungus, *Paecilomyces varioti*, in continuous culture to make protein, sold as Pekilo™. In Britain a similar process uses flour waste from flour making to grow a fungus, a species of *Fusarium*. The fungus is grown as a submerged culture in a thick mixture of carbohydrate-containing substrate together with other nutrients. Bubbling air through helps to circulate the mix. The mycelium is harvested and processed to make fibres, which are pressed together to make a material called mycoprotein. The mycoprotein has a high protein and fibre content but no cholesterol. The texture is made to be like meat chunks and it can be flavoured. It is used in pies and other products under the name Quorn™.

4.8 FOOD AND BIOTECHNOLOGY

The food industry, like others, has been affected by biotechnology. In particular, microbial enzymes are used in several processes. Many pre-prepared and processed foods need a strong cheese flavour at low cost to the manufacturer. **Lipase** enzymes modify immature cheeses to release free fatty acids that produce more intense flavours. Similar enzymes acting on milk fat produce buttery flavours for inclusion in a range of products. **Lactase** made by microbes is used to reduce the lactose content of milk for people with lactose intolerance. You can find out more about this process and the use of pectinase to clarify fruit juices in Chapter 6. Other enzymes are used to turn whey into a sweet syrup for alcoholic drinks and confectionery manufacture.

BST is **bovine somatotrophin,** a natural hormone made by cows. There is a link between the normal level found in a cow and the amount of milk she gives during lactation: BST promotes the release of a hormone, IGF-1, that stimulates glands in the cow's udder. High yielding cows have naturally high levels of BST. Genetically modified *E. coli* are used to make recombinant BST, which is used in the USA to boost milk production in dairy herds. It is not currently used in European dairy farming. Though BST is not biologically active in humans, there are animal welfare concerns, as cows treated with the hormone seem more likely to suffer other health problems.

After an animal has been slaughtered, natural breakdown processes in its cells makes meat more tender. This affects the ways in which meat can be cooked. **Meat tenderiser** is a protease preparation used to enhance the natural breakdown of tissue, making tougher meat more suitable for quick cooking methods. Papain, extracted from the papaya, is the main source of meat tenderiser. A number of proteases made by microorganisms are also available to treat meat and meat products in processed foods. For example, a protease from *B. subtilis* liquefies gelatin, whereas a protease from *A. oryzae* degrades protein in biscuit dough to make crisper biscuits.

QUESTIONS

4.7 Make a summary of the use of a **named** microorganism in food production.

4.8 Silage is made when the leaves of plants such as sugar beet, grass, maize and lucerne are compacted in silos with molasses and some mineral acid. What is the likely purpose of (i) the molasses, (ii) the mineral acid? Which sorts of bacteria are likely to grow in the silage?

4.9 Reread Section 4.5 and construct a flow diagram for cheese making.

4.10 Construct a chart of the food processes described in detail and for each list the microorganism used, the substrate used by the microorganism, say whether the process is a continuous culture or a batch process, and if it needs aerobic or anaerobic conditions.

4.11 Look up the efficiency of energy transfer through a food chain to explain why it is more sensible to use plant proteins directly rather than use them to feed animals.

Microorganisms can grow and multiply in all sorts of environments, which causes food spoilage problems. They change flavours and textures, and may produce toxic materials. The microorganisms themselves may cause human disease. Foods have to be stored, processed, and handled in ways that minimise food spoilage and reduce health hazards.

Unprocessed foods have protective outer layers, or contain anti-microbial chemicals such as lecithins or lysozyme. Microorganisms can enter and grow if the surface layers are damaged. A 'window of opportunity' is presented when foods undergo **autolysis**, that is breakdown due to their own enzyme action, during storage or chopping and mincing foods. It is also an opportunity for microorganisms to get into the food. Spoilage is rapid when environmental conditions are favourable. The variety of spoilage organisms and their effects are shown in Table 4.2. High protein foods are affected by a number of organisms, *Pseudomonas, Micrococcus, Bacillus,* and *Proteus* particularly. Some are psychrophilic (look it up in Chapter 3), and grow on meat in cold storage, making it go green or black. This may not be harmful but is aesthetically unacceptable. Carbohydrate-rich foods such as fruit, bread, jam, and fruit juices are spoiled by fungi such as *Penicillium* and yeasts. Other fungi and bacteria cause soft rots on fruit and vegetables.

TABLE 4.2 Some specific spoilage organisms

FOOD	DAMAGE	ORGANISMS
wine, beer	souring due to ethanol oxidised to acetic acid	*Acetobacter*
peanuts, cereals, fruit, dried foods, bread	aflatoxin produced causing food poisoning, mycelial growth through food	*Aspergillus* spp
pasteurised milk	'bitty milk' from lecithin breakdown	*Bacillus cereus*
most foods	green rot at low temperature	*Cladosporium*
pickles	blackening, as SO_4^{2-} reduced	*Desulphovibrio*
onions, potatoes	soft rot due to pectinase production	*Erwinia*
dairy products	felty red, yellow, orange, colonies	*Geotrichum*
citrus fruits, cheese	soft rot, green growth	*Penicillium*
bread, fruit, vegetables	visible mycelial growth	*Rhizopus, Penicillium*

Pathogenic organisms cause disease and are a major problem. They are most likely to be found in food derived from animals, for example meat, eggs, dairy produce and shellfish. Many of the organisms grow in animal intestines and enter food during handling. Shellfish such as mussels feed by filtering the water they live in for food particles, and prawns are scavengers. They may carry faecal microorganisms from sewage contaminating the water they live in. Great care has to be taken maintaining water quality and in cleaning shellfish after collecting.

Food poisoning

Most food poisoning is a result of microbial contamination of food. The symptoms are vomiting and diarrhoea, and the resulting dehydration can be life threatening.

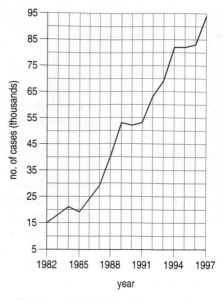

FIG 4.7 The incidence of food poisoning in England and Wales 1982–1997.

There has been a steady increase in the number of food poisoning infections, see Figure 4.7. Many different bacteria, protozoa and viruses cause food poisoning, but most cases are the result of just a few species of bacteria (see Table 4.3). Some infections are due to microorganisms multiplying inside the body; others are the result of **enterotoxins** produced by microorganisms. Species of *Salmonella* or *Campylobacter* are the main culprits.

TABLE 4.3 The important food poisoning organisms

ORGANISM	EUROPEAN SOURCES
Staphylococcus aureus	cooked meat, salted meat, milk based foods
Salmonella enteriditis phage type 4	eggs
Salmonella typhimurium	poultry, incompletely cooked food
Campylobacter jejuni	meat, poultry, handling animals
Listeria monocytogenes	milk products such as soft cheese, pate
Escherichia coli O157	contaminated raw beef

Salmonella multiply in the gut and damage the lining. Poultry are common sources of *Salmonella* and *Campylobacter*. Only a small proportion of chickens within a flock carry *Salmonella*, but cross-contamination occurs during processing. Chickens are chilled very quickly after death to slow microbial growth. You can read more about how *Salmonella* and its relatives cause food poisoning in Chapter 8.

Other food poisoning organisms growing in the gut produce a toxin that affects cells lining the gut. The toxin produced by *E. coli* O157 enables the bacteria to enter and destroy gut cells. *Staphylococcus aureus* is also a toxin producer. It is a skin organism that grows in cuts, boils and other skin inflammations. When it gets into food it grows and produces the toxin that causes the food poisoning symptoms when the food is eaten. People involved in food preparation have to wear plasters over cuts on their hands.

SPOTLIGHT

Listeriosis

Listeria monocytogenes is a bacterium causing listeriosis, a form of food poisoning. People with reduced immune system activity are particularly vulnerable, and it has been linked to miscarriage in pregnant women, who are advised to avoid high risk foods. The bacterium occurs naturally in our everyday environment – drains, floors, moisture, stagnant water, food-processing equipment, and on farms. It can contaminate a variety of raw and processed foods; for example, bacteria in milk contaminate cottage and soft cheeses made from it. Vegetables and raw and cold meats can be contaminated at the farm, as many farm animals are healthy carriers of the bacterium. The bacterium can also enter foods as they are being processed. It grows well at the temperatures normally found in the 'fridge and chiller cabinet, and in low oxygen environments, so great care has to be taken when preparing and storing pre-packed salads, dairy products and so on to avoid contamination at the packing stage. The bacterium is particularly resilient, surviving conditions that kill others – such as salty environments like cheese, and inside the white cells of the immune system (see Chapter 7). It can even survive hand washing, so special care is needed by food processors.

The most common sources of infection include cold meats and chicken, reheated stews, steak pies and incompletely cooked chickens. Food poisoning bacteria in uncooked food can pass to pre-cooked foods, and survive in foods that have not been thoroughly cooked. Heating a beefburger to 70 °C for 2 minutes is the minimum to make it safe to eat. Though it takes a very high dose of *Salmonella* to initiate an infection, relatively few *Campylobacter* are needed. In recent years *E. coli* O157:H7 has become more common as a food contaminant, mainly in minced beef products, or as a result of cross-contamination. This causes severe illness, diarrhoea with blood loss and renal failure if the bacteria and their toxin get into the bloodstream. The organism appears in animal faeces, which can also be a source of infection.

Vegetables are unlikely to carry pathogens unless they have been irrigated with untreated sewage, or handled in an unhygienic way. There is, however, a health hazard in stored plant material. The fungus *Aspergillus flavus* and related species can grow on moist stored grains and nuts, producing a toxin, **aflatoxin,** which has been found to cause human and animal deaths.

SPOTLIGHT

E. coli O157:H7

E. coli can produce enterotoxins that affect the cells lining the gut. There are several varieties of toxin so the strains producing each toxin have a designated serotype. *E. coli* O157:H7 strain produces an enterotoxin which affects protein synthesis in gut lining cells. It was first reported as causing haemorrhagic colitis in the early 1980s in the USA; the victims had been eating hamburgers. Other outbreaks followed in northern America, then further afield in other developed countries, causing thousands of infections and a few deaths. Most of the infections have been food-borne, mainly through hamburgers or minced meat, though there were some more surprising routes such as fresh cider, crisp flavouring and radish sprouts. This strain of bacteria gained importance in the UK in 1996 with an outbreak of food poisoning that claimed 17 lives and affected hundreds of others in Scotland. Though there had been cases of *E. coli* O157 infection before, there had not been such a severe outbreak. The bacteria were passed to the victims through cooked meats. A recent survey of dog faeces found that a high proportion carried this strain, which could offer another route of infection.

QUESTIONS

4.12 Use Figure 4.7 to describe the general trend of food poisoning cases.

4.13 Use the information in the passages above to produce a simple 10 point guide to avoiding food poisoning at home. Present your points in the form of an A4 sized poster.

4.10 PRESERVING FOODS

Food preservation aims to stop spoilage organisms growing in the food and to prevent food spoilage by autolysis. All the methods available affect the food; for example, heat treatment to kill bacteria alters the flavour, taste and texture of foods and may affect the nutritional quality. Some preservation methods kill bacteria, described as **bacteriocidal**; others merely inhibit growth, these are **bacteriostatic**. A few methods combine both bacteriocidal and bacteriostatic practices.

Canning

Preserving food in tin cans started over 150 years ago and revolutionised food storage, though steel is now used instead of tin. Food is packed in tins, heated, sealed so that no further organisms can enter, and subjected to sterilising temperatures to kill organisms within the tin. Using steam under pressure produces the high temperature necessary to sterilise the food. The higher the temperature the less time it takes to kill microorganisms. Even the most resistant spores will succumb after 20 minutes at 121 °C at pH 7. Sterilising time is shorter for acidic foods like tomatoes (pH 4) because acidity inhibits the growth of most spores. Chlorinated water is used to cool the tins. Tinned food lasts for a long time but the oxidation of fats in meat makes meats rancid. Tinned fruits keep less well because the acid content corrodes the tin. If the tin is dented, there is a chance that the seam welds could fail and microorganisms from the air or cooling water could enter and grow.

Freezing and refrigeration

Microorganisms grow slowly at low temperatures, and most stop growing below freezing point. Food is frozen by either passing it between two very cold plates or through a blast of cold air. Freeze-dried food is put in a vacuum chamber after freezing so that water sublimes off. At the time of freezing, food has its original complement of microorganisms plus any it has acquired during handling. A very small number of microorganisms continue to grow, though slowly, at freezer temperatures of –20 °C, including bacteria such as *Micrococcus* and fungi such as *Cladosporium*, *Penicillium* and *Monilinia*. Once food thaws, all viable bacteria begin to multiply and benefit from access to cells damaged by ice crystals. They reach high numbers quickly.

A domestic refrigerator holds food at +4 °C or lower. At this temperature microbial growth is slowed but not stopped, prolonging the life of perishable foods such as salads and milk for a few days.

FIG 4.8 Farmers may pasteurise their own milk, but usually it is done at the dairy before processing. Milk passes between metal plates heated to high temperatures then rapidly cooled.

Pasteurisation

Pasteurisation is used for foods that would be spoilt by temperatures needed for sterilisation. Many spoilage fungi and bacteria are killed if they are held at temperatures of around 60 °C for 30 minutes or more. In general, the higher the temperature the shorter the time needed to kill the microorganisms, but also the greater the effect of heat on the product. Spores and some highly resilient species can endure pasteurising temperatures and so pasteurised products will eventually 'go off'.

Milk is pasteurised by heating to 72 °C for 15 seconds then cooling rapidly (see Figure 4.8). This kills spoilage organisms, such as lactobacilli, and some disease-causing microorganisms that are transmitted through milk. *Mycobacterium tuberculosis*, which causes TB, was frequently carried in cows' milk. Milk sours after a few days because of the activity of streptococci and spore forming bacilli that survive pasteurisation. Other dairy foods such as liquid ice cream are pasteurised, as are fruit juices, wine and beer.

UHT milk is sterilised. It has superheated steam blown through it at temperatures of 135–160 °C for 1–2 seconds, which kills both spores and ordinary cells.

Low water potential

Some foods are preserved using salt or sugar. High sugar or salt concentrations draw water out of cells, and organisms find it difficult to grow in materials with such low water activity. Jams and conserves are 50–70% sugar and are boiled before packing into hot jars. This sterilises the jam, which is sealed to prevent further contamination. Once the jam is opened, bacteria still cannot grow, but some yeasts and fungi can tolerate the low water potential. Both salt and sugar absorb moisture from the atmosphere and this accelerates spoilage as the concentration of sugar or salt falls. Between 20 and 30% salt is used for salting and brining raw meat, raw fish and vegetables. There are some halophytic organisms that can tolerate these high concentrations and cause spoilage.

Drying

Dried food has a very low free water content, which reduces microbial growth. Dried cereals, grains and fruit, which have been dried by blowing hot air over them, can be kept for a long time. If the atmosphere becomes moist, the food absorbs moisture and the free water content rises, allowing microorganisms to grow.

Smoking

Smoking is another traditional method of preserving fish and meat, shown in Figure 4.9, though some food is salt cured first. Food is suspended over a fire and hot smoke permeates through it, which kills many microorganisms, denatures enzymes, and dries out the food. Many of the components of wood smoke are bacteriocidal or bacteriostatic, including cresols, organic acids and aldehydes such as formaldehyde. Spoilage can occur if smoked foods become moist.

FIG 4.9 Scottish Arbroath Smokies are fish which are split and smoked over a fire in a smokery to preserve and flavour them.

Preservatives

Preservatives are used to stop microbial growth, or natural breakdown, or to prevent chemical oxidations. Some of the processes already mentioned could be put in this category, for example smoking or salting food. Vinegar, often as pure ethanoic acid, preserves pickles, mayonnaise, sauces and coleslaw because the acidity inhibits microbial growth. Citric acid and sorbic acid can be used in the same way. Food preservatives and other additives are known by their EU code number, and are included on food labels. Newer additives have to undergo stringent safety testing. Table 4.4 shows some of the more common anti-microbial additives. Fruits and similar foods may be coated with biphenyl or o-phenyl phenolate to delay spoilage.

TABLE 4.4 Some common preservatives

EU CODE NUMBER	PRESERVATIVE	USES
E200	sorbic acid	dried fruits, low fat spread
E210	benzoic acid	foods made with fruit pulp
E211	benzoic acid salts	soft drinks
E220	sulphur dioxide	dried fruit, wines, pickles
E221–7	sulphur dioxide compounds	preserves vitamin C
E250–2	sodium nitrates and nitrites	meat
E260	ethanoic (acetic) acid	pickles, sauces, mayonnaise
E270	lactic acid	soft cheese products
E280	proprionic acid	
E296	malic acid	mincemeat
E297	fumaric acid	

Modified atmosphere packaging

Many spoilage organisms are aerobic, requiring oxygen for respiration. Modified atmosphere packaging replaces the air in pre-packaged fresh foods with a gas mixture which is low in oxygen, such as 35% carbon dioxide/65% nitrogen used for meats with a high fat content. This reduces changes due to chemical oxidation as well as slowing microbial growth. It leaves the colour, texture and flavour of the food unchanged. A few bacteria continue to grow in modified atmospheres, but the high carbon dioxide concentration slows most bacteria, particularly Gram-negative bacteria such as *Pseudomonas* and *Enterobacteriacae*. Modified atmosphere packaging is used for meat, poultry, prepared fruit, vegetables and salads, as well as fresh pasta.

Radiation

Irradiation inhibits the growth of microorganisms in food by affecting their nucleic acid and enzymes within the cells. Irradiation has long been used to sterilise and preserve culture media, medical supplies and drugs but some countries do not allow it for food preservation. Where it is allowed, it is used particularly to inhibit microorganisms in chickens and shellfish, and to stop sprouting in grains and vegetables.

CASE STUDY

Use of irradiation in food preservation

Foods can be irradiated by gamma radiation from the isotope ^{60}cobalt, which is cheap and relatively easy to handle, and by ionising energy from electron beams. The dose required to kill organisms varies: 10 kGy for bacterial and fungal cells; spores take 10–50 kGy. Viruses are very resistant, needing 30–50 kGy. In contrast, organisms as complex as humans are killed with doses of only 0.06–0.1 kGy, which causes damage to bone marrow cells. Generally a dose of 45 kGy is regarded as a sterilising dose. Radiation can be used on foods as a means of sterilising them, or as a partial steriliser in which spoilage organisms are destroyed but others may still be present.

Gamma radiation has other effects which extend the storage time of fruit and vegetables. 20–150 Gy inhibits the sprouting of onions and potatoes after harvest, and delayed ripening extends the market life of fruit by up to 20 days. Up to 25% of stored grain is lost by insect damage caused by mites and beetles. Grain is treated with organophosphate pesticides to kill them but the insects are becoming resistant. In Canada, levels of 750 Gy have been approved for killing adult insects, larvae, eggs and pupae in grain. Unlike insecticides, irradiation leaves no residues nor is there reinfestation from insect eggs. Irradiation can kill parasites in meat and fish too.

Irradiation to preserve food is commended by the World Health Organisation and by the Food and Agriculture Organisation among others. It was approved in the UK in 1986 but has not yet come into common use. However, world wide 40 countries have approved irradiation for over 60 food products; irradiated food even goes on the space shuttle. In the UK there are seven categories for which irradiation could be used up to specified doses. These include

- poultry, fish and shellfish, to reduce food poisoning organisms,

- potatoes and onions and similar bulbs and tubers to reduce sprouting,

- fruits and vegetables to reduce spoilage organisms,

- cereals to reduce insect pests,

- spices and condiments to improve overall microbiological safety.

Only spices and condiments are irradiated at the time of writing. Products must be labelled and there is legislation for the irradiating establishments.

Irradiation could be useful for controlling food-borne illness because food can be treated after packaging and sealing, preventing any further contamination. This is particularly important with the increase in 'ready-meals' and the presence of *Listeria* in pre-prepared foods. Its main advantage is that it can reduce losses of food and preserve high quality food, but it will not improve poor quality food. It is the only method available at the moment for eliminating *Salmonella* from frozen meat and bulk animal feeds.

Concerns centre around several issues:

■ Is there a risk of some spores surviving to cause spoilage that a consumer may not recognise? This very minute risk will depend on the dose used, the quality of food manufacturing practice, and the source of the food.

■ Is there a loss of nutrients in irradiated food? In fact, this is similar to the loss in foods treated with more conventional forms of heat treatment.

■ Does irradiation make food radioactive? Radioactive sources with energy levels below 10 MeV must be used to ensure there is no induced radioactivity. When ^{60}cobalt is used, induced radiation can't be detected.

■ Does the radiation affect chemicals present in the food? No, nor is there conversion to toxic compounds if doses below 50 kGy are used.

The disadvantages of irradiation for food preservation are

■ it is expensive,

■ high fat foods may develop off-odours and tastes because rancidity is hastened,

■ meat flavours may change, but this is minimised by carrying out irradiation at chiller or frozen temperatures,

■ methods of detecting whether food has been irradiated are recent developments, just becoming sufficiently reliable for routine monitoring.

> ### *More information*
>
> ■ World Health Organisation: www.who.int
>
> ■ Food and Agriculture Organisation: www.fao.org

QUESTIONS

4.14 What is the difference between the terms 'bacteriocidal' and 'bacteriostatic'?

4.15 Review the list of preservation methods and list them under three headings: 'bacteriocidal', 'bacteriostatic' and 'both'.

4.16 Why is it considered unwise to (a) buy dented tins, (b) refreeze food which has already been frozen and defrosted once?

4.17 Many large-scale caterers are adopting a 'cook–chill' technique of preparing meals in advance followed by reheating at the point of use. Why is this technique thought to be better than cooking then holding in warming trolleys?

4.18 How do fungi spoil food? How may such spoilage be reduced?

4.19 Access the Institute of Food Science and Technology's website to find out the latest information on health hazards in food.

> ### *More information*
>
> ■ Institute of Food Science and Technology: www.ifst.org

4.20 Review the chapter and examine the case study on the use of radiation for food preservation.
   (a) List the points for and the points against the use of radiation on food.
   (b) You are a member of the Meat Pie Manufacturers Association (MPMA) in favour of radiation and are to present evidence to a Government committee considering legislation on the use of radiation on foods. State your case.
   (c) You are now a member of Clurshire Consumers Group (CCG). What safeguards would you expect the Government to include to protect the general public?

   You may prefer to do this question as a group activity. Elect a chairperson, let two people prepare the case for the MPMA and two for the CCG. Everyone else is a Member of Parliament on the committee. What decision did you come to?

4.21 Bovine somatotrophin (BST) was used in anonymous trials in Great Britain to monitor the effect on milk yield. Many people were upset at not knowing whether their milk had come from cows treated with BST. Why do you think people were worried? Would it have made any difference to the trials if people did know which herds were being treated?

Exam questions

4.22 Malting is a process used in the brewing of beer. Barley grains are germinated. During this process enzymes called amylases convert the starch they contain to glucose and other sugars. After malting, the malt is dried in a kiln, ground up and mixed with warm water. Hops are added to give flavour and this mixture is heated to boiling. The resulting wort is cooled by blowing air through it and a yeast culture is added. At first, the yeast cells grow and divide but do not produce ethanol. Later, they stop dividing and start producing ethanol.

(a) What type of carbohydrate is (i) starch; (ii) glucose? (2)

(b) Explain why malting is necessary. (2)

(c) Suggest two reasons for boiling the wort. (2)

(d) Explain why there is a delay before the yeast produces ethanol. (2)

AQA Biology, March 1999, Paper BY06, Q. 3

4.23 Yoghurt is a semi-liquid product of fermented milk, produced as a result of a mutually beneficial relationship between two species of bacteria, A and B.

(a) Name the **two** species of bacteria involved in the fermentation. (2)

The graph shows the growth of these bacteria and the pH of the medium during fermentation.

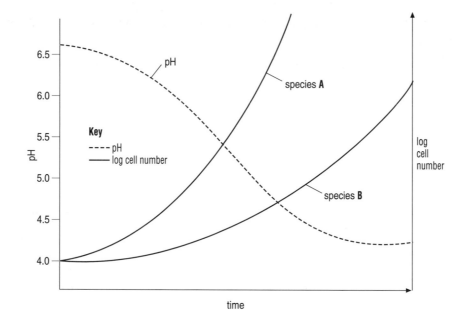

(b) With reference to the graph (i) describe the growth of species **A** and **B**, in relation to the production of yoghurt; (ii) explain the mutually beneficial relationship between the two organisms. (2)

(c) Explain why the milk is heated to between 85 °C and 95 °C and then cooled before the starter culture is added. (2)

(d) Name the fermentable energy source in the milk. (1)

(e) (i) Name **two** substances which give natural yoghurt its flavour.

(ii) State and explain a treatment of the yoghurt after the fermentation is completed. (2)

OCR Sciences, June 1999, Paper 4806, Q. A1

SUMMARY

Microorganisms live on the surface of animals and plants and enter food as it is prepared. Microorganisms also live in the gut and in surface ducts of animals and can enter food from these sources.

The processes that microorganisms use to obtain nutrients from food sources bring about chemical and physical changes in food. Using microorganisms to process foods involves the manipulation of growth conditions to select for the growth of the desired organisms and to select against the growth of others.

Lactic acid bacteria require high levels of nutrients and produce lactic acid as a metabolite, which produces physical changes in the food. Lactic acid bacteria use lactose in milk and the lactic acid affects milk protein structure to make cheese and yoghurt.

Yeast uses sugar to make ethanol in beer and wine and carbon dioxide which raises bread.

Aspergillus converts proteins in beans to peptides in soy sauce production.

Microbial enzymes such as lactase are used to change food constituents in particular products.

Microorganisms growing on food can spoil it or generate toxins. Food poisoning is caused by a number of bacteria found in animal gut.

Bacteriocidal food preservation methods kill organisms but bacteriostatic methods reduce their growth rate.

There is controversy over the use of radiation as a method of preservation.

5 CLEANING UP WITH MICROORGANISMS

LEARNING OBJECTIVES

After studying this chapter you should be able to:

① understand the problems caused by inadequate waste disposal

② describe ways in which microorganisms are used to provide clean water

③ explain how microorganisms are used to convert waste materials into useful products

④ describe how microorganisms can be used to reduce environmental pollution.

5.1 PROBLEMS GENERATED BY WASTE DISPOSAL

Dead material and wastes from animals and plants are usually degraded into smaller molecules by living organisms. The same organisms degrade many wastes from human activities too. However, some are not easily broken down, including many compounds synthesised within the last century, and they accumulate to offensive levels, or pose human health hazards, or are toxic to other living things.

Domestic and industrial wastes are buried, incinerated or dumped in rivers and seas. We know that this is not a satisfactory long-term solution. Dumping is cheap but creates new problems as harmful compounds accumulate and some fairly harmless compounds may be converted into more harmful materials. Mercury, lead and other metals, and some pesticides are taken up by living things from low levels in the environment. These pollutants accumulate within the organisms and are passed on to other animals higher in a food chain, where the concentration may be harmful (**bioaccumulation**).

A second issue is the spread of disease by wastes. Human health has improved significantly over the last 150 years and there are far fewer deaths from infectious disease. The single most important factor contributing to this is clean water, uncontaminated by harmful organisms carried in sewage. Where water supplies are contaminated by sewage, tens of thousands of people die each day from water-linked diseases including typhoid, diarrhoea, cholera, guinea worm and river blindness. Many more are ill, reducing the workforce and straining resources.

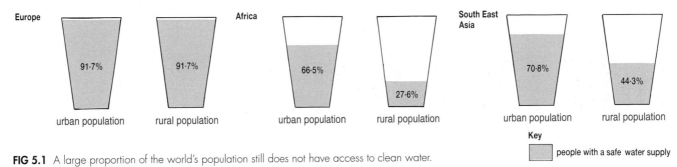

FIG 5.1 A large proportion of the world's population still does not have access to clean water.

The same microorganisms that decay natural wastes are widely used to treat sewage and increasingly to deal with pollutants and harmful chemicals. They are also harnessed to make useful products from wastes and to clean up the environment.

5.2 NATURAL WATER

Water in rivers and lakes carries silt, grit, dissolved salts, vegetation fragments, and soil microorganisms brought by rainwater run-off. Open water has few nutrients, which limits the number of microorganisms that can grow there, but nitrogen-fixing photosynthesisers such as cyanobacteria thrive in the warmer surface waters. There are nutrients trapped in crevices between dead wood and stones on the beds of lakes and rivers. Here, microorganisms secrete enzymes into their immediate surroundings to break down organic material and absorb the soluble molecules produced. Enzymes work more effectively in crevices because they are not dispersed. These microorganisms include soil bacteria such as *Streptomyces* and *Bacillus* species washed into the water by rain. Very deep or very turbid waters may have very low oxygen concentrations near the bottom; only microorganisms that can respire anaerobically live there.

5.3 POLLUTED WATER

Pollutants change both the physical nature of the water and the balance of living organisms in it for many kilometres downstream of the entry point. Pollutants come from commercial processes and factories, farms, towns and settlements. There are five important ways in which pollutants can affect watercourses.

Organic waste

Organic wastes include food processing wastes, urine and faeces in sewage and farm wastes. Farmers store grass in silage clamps to feed cows in the winter; the grass ferments producing organic acids which leach into water draining through the clamp. This is more polluting than waste from intensively reared animals, unless it is treated. Organic matter allows microorganisms which are not normally found in water to flourish, and populations of all bacteria soar from 10^2 to 10^6 per cm^3. Organic wastes decrease dissolved oxygen as the waste is oxidised to carbon dioxide, nitrate, phosphate and other compounds. Algae in surface waters grow rapidly because of the increased nitrate concentration, leading to eutrophication, detailed in Section 2.4. Anaerobic organisms flourish and generate acids, sulphides and amines, which affect other aquatic organisms. Organic chemicals can also cause unpleasant smells or have toxic effects on the water inhabitants.

Inorganic wastes

Inorganic wastes include fertilisers and salts as well as ions such as mercury, cadmium and lead used in industrial processing. Nitrate and phosphate in crop fertilisers leach into watercourses and disturb the ecological balance of water organisms. Nitrates are also released by sewage treatment. Plants and microorganisms take up metal ions in aquatic ecosystems, which bioaccumulate to toxic levels higher in the food chain.

Particles

Small particles and silt suspended in the water from mining, quarrying and other industries block the passage of light through the water, reducing photosynthesis in aquatic plants. The particles also clog gills and delicate body structures of aquatic animals. Sediments make it difficult for river inhabitants to hold on to the substrate, so they are washed away.

SPOTLIGHT

Introducing Public Health Authorities in Britain

At the beginning of the 19th century there was no regular system of rubbish disposal in Britain. Usually waste water and sewage was tipped out onto land or into drains that went into the local brook, which also supplied drinking water. Life expectancy was about 40 years and sickness and disease were rife. In 1831, after an outbreak of cholera, a Central Board of Health was set up which recommended providing a system of public health authorities. The authorities were set up but were rather ineffectual. Investigations into major outbreaks of disease in 1838 found that the physical environment of the poor encouraged sickness. A report on conditions stressed the economic cost of sickness in the 'labouring classes', and the effects of bad housing and sanitation on health.

Little was done but by 1848 the first Public Health Act was passed and Medical Officers of Health were appointed. Houses had to have a water closet, a cesspit or an ash pit. Weekly sickness returns were compiled and information was systematically reported. At the time people thought infectious diseases were caused by miasmas in the air and the main aim was to provide clean healthy surroundings where miasmas did not develop. Fortunately the sanitary engineering systems were effective against the microorganisms causing disease.

In the middle of the century Dr Snow investigated a cholera outbreak, and although he did not know the cause, he established its source as contaminated water supplies. The need for clean water was recognised.

Harmful microorganisms

Disease-causing organisms enter water with untreated sewage. They include cholera and typhoid bacteria, polio virus and the eggs of parasitic worms. The organisms are passed on when contaminated water is used for drinking, washing or preparing food. Significant numbers of enterobacteria from the gut can survive in warmer water.

Heat

Many industries use water as a coolant; cooling water gains heat in the process and is discharged into rivers where it adversely affects aquatic life. Though microorganisms can grow in a wide range of temperatures, most cannot survive the sudden changes in temperature that can occur near cooling water outlets.

QUESTIONS

5.1 Suggest four sources of water contaminants.

5.2 There are few microorganisms in open waters. Why do the numbers increase if untreated sewage is discharged into the water?

5.3 Write definitions of the following: heterotrophic microorganisms; turbid; anaerobic respiration.

5.4 Review Section 2.4. Outline what happens in eutrophication.

5.5 Read the information about the formation of Public Health Authorities then answer the following questions:
  (i) What is meant by 'the economic cost of sickness'?
  (ii) What is a miasma? (line 15)
  (iii) Why didn't people recognise that microorganisms caused disease?
  (iv) How and why do you think the regulations that each house should have a water closet (WC), a cesspit or an ash pit led to improvements in the health of urban inhabitants?

5.4 MEASURING WATER QUALITY

Several factors are used to measure water quality in rivers and at bathing beaches. The most important are

■ the amount of organic material present,
■ the turbidity of the water,
■ the numbers of particular microorganisms.

The amount of organic material is measured indirectly. The oxygen content of a water sample is measured. The sample is sealed and kept for 5 days at a standard temperature, then the oxygen content is remeasured. During that time the oxygen content will have fallen because of oxidation of organic material present. The amount of oxygen used is called the **Biochemical Oxygen Demand** or **BOD**, and is calculated in milligrams per litre. Generally the higher the oxygen consumption the more organic material is present.

Turbidity is due to colouring matter in the water and suspended particles. Turbidity is measured by the scattering of light as it passes through the sample. Particles are filtered out and weighed, recorded as milligrams **suspended solids** per litre.

Monitoring harmful organisms is far more difficult. The pathogenic organisms from a small number of infected people are well dispersed in the huge volume of bacteria-laden effluent produced by a community. Looking for these organisms is like looking for a needle in a haystack. However, untreated sewage carries large numbers of relatively harmless organisms called **faecal coliforms** (the *E. coli* -like bacteria that live in the gut of animals and man) which can be detected more easily. These organisms are called **indicator organisms,** because their presence indicates that there are untreated faeces in the water, possibly carrying pathogenic organisms. Their numbers are monitored to see if untreated sewage is entering.

Directives govern the maximum permitted levels of BOD, suspended solids and faecal coliforms, and water should comply with these almost every time it is sampled. There are more stringent guidelines, which should be the target for good water quality. Table 5.1 illustrates some of these guidelines. Other factors are also monitored, so river water will be assessed on pH, temperature, nitrate and ammonia content. Ammonia indicates potential sewage or livestock effluent entering the water.

TABLE 5.1 Water quality directives

BATHING WATERS DIRECTIVE

Maximum coliforms permitted

95% of samples should have not more than
10,000 total coliforms per 100 ml, 2,000 faecal coliforms per 100 ml

Guideline coliform and faecal streptococci standards

80% of samples should have not more than
500 total coliforms per 100 ml, 100 faecal coliforms per 100 ml

90% of samples should have not more than
100 faecal streptococci per 100 ml

RIVER WATER QUALITY

A grade A river will
be at least 80% saturated with dissolve oxygen
have a BOD of not more than 2.5 mg per litre
have an ammonia content of not more than 0.25 mg N per litre

Some water-borne diseases are caused by viruses, which are difficult to detect. Any viruses in a water sample are concentrated on a membrane filter and grown in cell culture. They are detected using specific antibodies (Chapter 6 tells you more about using antibodies to identify substances). This takes days and works best when there are large amounts of virus present. Quick tests using gene probes are being developed. They consist of a piece of DNA or RNA that corresponds to a stretch of virus nucleic acid. The probe will bind to viral nucleic acid present in the sample.

QUESTIONS

5.6 List three factors used to measure water quality.

5.7 Table 5.2 gives readings for the BOD and suspended solids in the effluent from four factories and two farms.

TABLE 5.2

UNIT	VOLUME OF EFFLUENT PER WEEK (1000 dm³)	AVERAGE READINGS (mg per dm³)	
		BOD	suspended solids
factory A	18	28	55
factory B	3	20	133
factory C	11	115	30
farm D	0.5	89	40
factory E	6	18	28
farm F	1	33	35

(a) Which of these concerns has effluent that exceeds the maximum permitted levels?

(b) Each of these farms and factories produces a different volume of effluent. Calculate the total suspended solids for each concern. Which contributes most to suspended solids pollution? Which causes the greatest BOD problem?

(c) Which of the factories could be a food factory and which a cement works?

(d) If one of the factories produces a hot water effluent, what effect will this have on decomposer microorganisms?

(e) Farm slurry is a very liquid mixture of animal faeces. Which of the farms could have slurry leaking into its drains?

5.8 Visit the Environment Agency website to find out the quality of water near you.

> **More information**
>
> ■ The Environment Agency: www.environment-agency.gov.uk

5.5 HOW WATER IS PURIFIED

In Britain, water companies are responsible for the supply of good quality water and waste water treatment. They are regulated by government agencies including the Environment Agency. The demand for clean water is huge: in just one region over 8 million people use 1.9 billion litres of clean water each day in homes and industry. Industries use most of their water straight from rivers, and it needs treatment before it can go back, so in the same region 2.7 billion litres of waste water including nearly 3 million litres of sewage are treated each day.

The supply

Water comes from three main sources:

■ **Boreholes** are deep shafts sunk into the underlying rock to draw high quality water from the water table. Water users have their own wells and boreholes.

■ **Reservoirs** often have high quality water but it is usually treated before distribution.

■ **River water** needs treatment because it contains undesirable materials.

Water is treated so that it has no suspended material or colour, its taste is acceptable and it does not smell unpleasant. There must be no harmful organisms nor an excessive amount of salts, which could deposit in pipework. Most purification techniques mimic the natural mineralisation process that occurs in soil and large bodies of water (see Section 5.7).

Treating the water

There are several approaches that are often used in combinations before water is distributed.

Filters

Filters mimic the processes that occur as water percolates through soil, and at the bottom of streams. The filter is a bed of carefully selected sizes of sand particles on gravel. Suspended solids are trapped as water passes through the filter. In slow filters, microorganisms coating the filter bed particles can degrade some organic material. Harmful microorganisms cannot compete with soil microorganisms and are likely to die in the low temperature. High quality water drains out from the bottom of the filter.

Reservoirs

Particles in the water of a reservoir settle to the bottom. Organic matter is oxidised rapidly by microorganisms, and the low temperatures do not favour harmful microorganisms. Colouring matter, such as tannins from peat, in the water is gradually bleached out and hydrogen carbonates are converted to carbonates, which settle.

Purification plants

River water has so many impurities that it needs more thorough treatment. Mesh screens remove debris from the water, which may be aerated to remove smelly gases. Chemicals such as aluminium sulphate are added to coagulate suspended solids, making a **floc**, which also traps some colouring matter and bacteria. The floc is drained off to a sludge lagoon. Water hardness is adjusted by salt precipitation or by ion exchange methods. Chlorine is added to kill any remaining microorganisms. The water is then passed through a filter bed. Rapid gravity filters, shown in Figure 5.2, are used in areas of high demand. These can deal with large volumes but are less effective than slow filters at removing tastes, colours and bacteria.

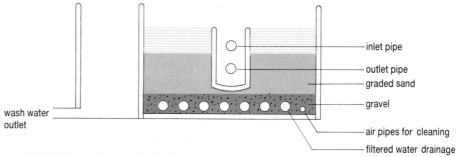

FIG 5.2 A rapid gravity filter bed.

Water in some areas may need extra treatment to cope with specific problems. Ion exchange is used to reduce high nitrate concentrations from fertiliser in some agricultural areas. Water pH is adjusted to close to neutral because acid waters corrode pipework and alkaline water leaves deposits. Fluoride salts are added to the water in most areas. Further chlorine reduces microbial activity in the pipelines and acts as a safeguard against minor leaks into the water supply.

QUESTIONS

5.9 What are the main sources of water for drinking?

5.10 Why does leaving water to stand in reservoirs result in fewer microorganisms in the water?

5.11 Explain how filters can be used to obtain pure water supplies.

5.12 Construct a flow chart of the processes used to purify river water.

5.6 DEALING WITH WASTE WATER

Waste water drains into sewers and consists of 99% water and 1% a complex mix of materials from a variety of sources. The largest water user is the electricity generating industry – a water-cooled power station can use over 200 million litres of water every hour. Chemical industries and engineering also use large amounts – nearly half a million litres to make a small car. Drains from industrial premises carry rainwater and materials from their activities. These include metals and inorganic ions from some industries, oils, greases, solvents, and food waste from food factories, although all have to abide by strict discharge limits. Storm water drains take rainwater, debris and grit from the streets into sewers.

If you live in Britain you are responsible for about 130 litres of effluent from your home each day. 30 litres from the kitchen carries detergents, disinfectants, bleaches and scouring agents, together with food waste. 35 litres of bath water carries grease, bath oils and soap. And each flush of the toilet uses 9 litres to carry away faeces, urine and toilet paper; this is the part referred to as sewage.

Altogether about two-thirds of the material in waste water is organic. It is broken up quickly and is carried as a suspension. Effluent treatments reduce the number of harmful organisms, remove offensive material, and degrade organic compounds to simpler chemicals such as nitrates, phosphates and carbon dioxide.

USING MICROORGANISMS TO CLEAN UP WATER

5.7

A waste water, or effluent, treatment plant harnesses natural microbial activity and provides the best possible conditions for their growth, and hence the efficient breakdown of organic material. The microbes secrete enzymes that degrade carbohydrates, proteins, amino acids, nucleic acids, lignin, cellulose and lipids. The treatment plant is engineered to ensure the high oxygen concentration needed to degrade organic material. Organic material is broken down to carbon dioxide, which is released to the atmosphere, and nitrate, sulphate and phosphate ions dissolved in the water. This process is called **mineralisation**. During the process a large amount of sediment and sludge accumulates, which is decomposed anaerobically making methane, a useful product. Harmful organisms die in the process. WHO guidelines ask that there should be no *E. coli* or thermotolerant coliform bacteria in any sample of water entering the distribution system. A waste water treatment plant can be seen in Figure 5.3.

FIG 5.3 A birdseye view of a waste water treatment plant. Identify as many of the structures as you can; the circles are trickle filters.

What happens in a sewage works

After screening to remove large debris, effluent is left in a grit settlement tank for a few hours for grit and stones to settle out. Some organic materials flocculate and settle making primary sludge. The waste water then passes through a trickling filter or an activated sludge treatment tank for secondary treatment.

Trickling filters

A trickling filter bed is made of carefully graded stones, grit and clinker. A complex ecosystem including bacteria, fungi and protozoa develops within it. Waste water is slowly sprayed over the top and microorganisms degrade organic matter as it trickles through the bed. The microorganisms are eaten in turn by predatory protozoa and insect larvae. Most matter has been mineralised by the time the water reaches the bottom.

Activated sludge

Waste water enters a large tank where it is vigorously aerated. Aeration mixes the contents as well as ensuring high dissolved oxygen concentrations. A starter culture of a microbial mixture of organisms called the **zoogloea** is inoculated into the water. *Zoogloea ramigera* is a zoogloeal species that secretes a gum which flocculates particles together. Colonies of organisms, including aerobic bacteria and protozoa, work within the floc breaking down organic matter. The flocculated particles settle out at a later stage. It takes about 8 hours for the BOD to be reduced by about 90%. A portion of the zoogloea is siphoned off when the water leaves the tank to inoculate the next batch of sewage. *Zoogloea* is sensitive to certain contaminants, so when water contaminated with, for example, heavy metal ions arrives, it has to be held back and diluted with ordinary sewage to avoid killing the organisms. Foaming detergents cause problems too because they reduce the movement of oxygen from the air into the water.

The water is good quality after secondary treatment but in many areas there is a tertiary level of treatment before the water is discharged into a river. Any remaining particles or zoogloea are settled out in a secondary sedimentation tank before a final filter in shallow gravel beds or passage through reed beds. The nitrate content of the water may also need to be lowered before it can be discharged.

Sediment from the primary and secondary sedimentation tanks is also treated. It can be dried or burnt but these waste a valuable resource. Sediment goes into large warm tanks called **digesters** where anaerobic microorganisms such as clostridia ferment it to produce acetate, other reduced carbon compounds and hydrogen. These are then used by methanogenic bacteria which make methane that can be used for power. Very few harmful organisms survive the treatments so it is usually safe to dry any remaining sediment for use as fertiliser or landfill. In industrial areas levels of metal contamination may prevent its use on agricultural land.

Other ways to treat sewage

In places where the population is too scattered for central sewage works, or water is in short supply, people use alternative sewage treatments. The most common method is the use of pit latrines, which are concrete-lined to prevent the contents from contaminating ground water supplies and designed to reduce the entry of flies.

Digesters

The organic material in sewage is valuable as a source of fertiliser or energy so there are systems that make the most of it. Sewage can be used with food waste and animal manure in **digesters** to generate methane for cooking and power; more about this can be found in Section 5.9. The process generates temperatures that are high enough to kill most harmful microorganisms.

Crop irrigation

Domestic water and sewage has been used to irrigate crops for thousands of years. A well-managed system will also help to protect clean drinking water supplies and add humus, nitrates and phosphates to the soil, reducing the need for fertilisers. Higher numbers of organisms are acceptable in water used for crops such as trees and cotton but it is better for irrigation water to be free from harmful organisms to reduce risk to field workers, crop handlers, local people and eventually consumers. Care must be taken to prevent contact between waste water and any edible parts of the plants. Ideally it should be used for tall growing crops such as olive trees or tomatoes.

Typically sewage runs into a settlement tank first to make a sediment which can be used or burned; the waste water runs through reed beds. These are impermeable shallow channels filled with gravel and planted with reeds (*Phragmites*). These reeds conduct air into their roots, and microorganisms around the root system break down organic material in the water. Reed roots also take up nitrates produced by bacterial action. Water passing from the channels has few harmful organisms and is used to irrigate crops. The reeds are harvested as a separate crop. Figure 5.4 shows a reed bed purification system.

FIG 5.4 Reed roots in gravel form a habitat sheltering organisms that degrade organic material. The reeds use minerals released into the water and are a valuable crop in their own right.

Sewage lagoons

Sewage lagoons are widely used. Sewage is pumped into large open ponds where natural decay processes degrade organic material. The lagoons are shallow enough to ensure that there is enough oxygen diffusing in for aerobic breakdown. However, this is a slow method and some harmful organisms may survive in warm climates.

QUESTIONS

5.13 List three methods used to purify waste water and sewage.

5.14 What are the main advantages of using microorganisms to purify water?

5.15 Explain the meaning of the terms mineralisation, floc, digester.

5.16 What is the role of floc in sewage treatment?

5.17 List the major physical, microbiological and chemical changes in effluent as it passes through the treatments in a sewage works.

5.18 What useful products are made by anaerobic digestion of sewage sludge?

5.19 Name three organisms likely to be found in a trickle filter.

5.20 (a) Explain the main sources of pollutants found in river water.

(b) Table 5.3 shows the results of an analysis of samples of two different river waters, both suspected to have been polluted recently.

TABLE 5.3

RIVER SAMPLE	BOD $mg\ dm^{-3}$	NO^{3-} $mg\ dm^{-3}$	NH^{4+} $mg\ dm^{-3}$	COLIFORM CONTENT AS *E. coli*, PER 100 cm³
A	32	60	0.75	> 18 000
B	3	4	0.10	< 50

> ### More information
> - The Centre for Alternative Technology: www.cat.org.uk
> - Science Traveller International: www.scitrav.com/wwater

(i) Explain how each of the four tests could show water pollution.

(ii) Explain which of the river samples would be most suitable for domestic or industrial purposes.

5.21 You can research water treatment via websites explaining waste water purification. The Centre for Alternative Technology has research into composting toilets and using sewage to irrigate crops. The University of East London and the University of Florida have web tours of treatment works. Science Traveller International provides links to many other websites.

Many industrial processes generate unpleasant wastes that cause ecological and aesthetic problems when they enter water or soil. They may linger for decades or accumulate to concentrations that cause concern. Even worse, relatively stable chemicals may be degraded into toxic substances by soil and water microorganisms. Economic and health problems arising from oil spills and long-term pollution has encouraged research into using microorganisms to degrade toxic wastes into harmless chemicals. Removing pollutants is difficult and often expensive. Pollution is usually treated mechanically or chemically, using filters or chemical cleaning processes. It is becoming profitable to use microorganisms to change pollutants in one of three main ways:

- they can degrade the complete range of carbon compounds and most nitrogen-containing compounds;

- they may have useful indirect effects, such as changing the solubility of a pollutant when they release organic or other acids as a by-product of metabolism. There is even a strain of *Thiobacillus* that can grow on concrete contaminated with radioactivity, slowly dissolving it away with the acids it produces;

- they can take up metal ions against the concentration gradient, and tolerate higher concentrations than many plants.

Using organisms to clean up the environment is called **bioremediation**. Bioremediation is being used against wastes from metal processing, plastics and other industries, oil spills, mining spoil, herbicides and pesticides.

Oils and plastics

Hydrocarbons in crude oil are made into a range of oils, polymers and plastics such as PVC, polyethylene and polyurethane, which cause disposal and pollution problems. Waste lubricating oil from engines and plastic wrappers cause particular problems because soil organisms degrade them very slowly or not at all. However, there are specialised groups of microorganisms that can grow in fuel oils and lubricants if there is some moisture available. Members of over 20 different groups of bacteria can degrade hydrocarbons. Crude oil in oil fields carries these organisms but there is not enough water for their growth. The microorganisms grow when moisture enters the oil after it has been pumped up, or used in engines and machinery. They break down hydrocarbons, generating corrosive metabolites such as organic acids, sulphuric and nitric acid, and ammonia. Figure 5.5 illustrates the general pathway. The organisms are troublesome in fuel pipelines and engines but a genetically modified strain of *Pseudomonas* is efficient at cleaning oil spills. Other species are under investigation for cleaning up oil spills onto soil.

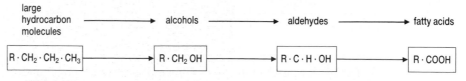

FIG 5.5 How hydrocarbons are oxidised by microorganisms.

Pseudomonas aeruginosa degrades plasticised vinyl quite quickly and a fungus, *Cladosporium resinae*, degrades paraffin-based fuels such as aircraft fuel. Plastics containing plasticisers are degraded by organisms that can metabolise the organic plasticiser molecules. Some packaging plastics have been specifically developed to be microbially degradable. Polyhydroxyalkanoic acids are made by the action of a species of *Alcaligenes* on carbohydrates and can be made into a polymer, Biopol, which is completely biodegradable. It is oxidised when in contact with normal soil organisms, even in low oxygen conditions such as landfill sites. Microorganisms such as *Pseudomonas* can also degrade water-based paints and polyurethane coatings, corrosion inhibitors in car radiators, adhesives and synthetic materials.

Disposing of old car tyres is a growing problem. Some are pulverised and recycled by mixing with new rubber for manufacturing, but the sulphur left by the original vulcanisation affects the quality of rubber made with some recycled material. Research is underway to find sulphur-metabolising bacteria to degrade sulphur compounds in old vulcanised rubber, making recycling rubber more effective.

Farming wastes

In some areas **nitrate** from agricultural fertilisers and effluent treatment plants entering water raises concentrations above the guideline of 50 mg per litre. The problem is worst with fertiliser applied to slow-growing winter crops, or if there is heavy rain after the fertiliser is applied, and when land is ploughed. Farmers in areas around sensitive boreholes are limited in how much nitrate fertiliser they can apply to their fields, which reduces nitrate entering the water at source. Contaminated water is passed through an ion exchange column that exchanges hydrogen carbonate ions for the nitrates. The nitrates are then washed out of the column for disposal. Nitrates in water are difficult to remove chemically on a large scale. Potentially, denitrifying bacteria could reduce the nitrate to nitrogen gas to release into the atmosphere, or microorganisms could be used to take up nitrate from polluted water.

Persistent herbicides and pesticides pose a different sort of problem under investigation. There are strains of *Arthrobacter* that can degrade phenylurea-based weedkillers; potentially they could be used to remove residues from soil.

Metals and industrial pollutants

Metal ions, such as mercury and cadmium, enter soil and water as a result of industrial processes, as diverse as paper-making and 'metal-bashing' industries. Such metals are toxic in quite small quantities and accumulate in food chains. The reclamation of unstable spoil tips or lagoons containing toxic waste materials poses another problem. These areas may be short of nitrates and other plant nutrients, and generally unsuitable for the growth of plants. Extracting low concentrations of metals from water is both physically and chemically difficult. Often it is not worth while, except for valuable metals such as gold. Usually the metals are precipitated from the water using chemicals and filtered out. The filtered sludge is toxic, expensive and awkward to dispose of. Increasing awareness of heavy metal pollution has led to pressure for better techniques for extracting metals from low concentrations in water.

FIG 5.6 The spoil tip at this Nottinghamshire colliery is being reclaimed and turned over to agriculture. The spoil heap needs stabilising and seeding to encourage the growth of tolerant pasture species.

Particular attention has been paid to microorganisms that take up and accumulate metal ions from dilute solutions. The uptake may be against a concentration gradient and the metal incorporated in a non-toxic form. Organisms such as the unicellular alga *Scenedesmus* and some fungi are very promising. Mycorrhizal fungi (those that live in a mutualistic relationship with plants) have been shown to take up metal ions from dried sewage sludge used as a soil conditioner. Some bacteria can even accumulate mercury without coming to too much harm. Other research focuses on using crop plants grown hydroponically to take up metals from waste water through their roots. The metals can be recycled after the crop is harvested and burned. Indian mustard, *Brassica juncea*, even tolerates the uptake of radionucleotides such as strontium.

Research has been done on using algae to extract precious metals from seawater and jewellery industry wastes. The development of photobioreactors where algae are grown in controlled conditions could make this more feasible. Microbial mats of cyanobacteria and other microorganisms with an organic matrix material such as grass clippings are very effective at degrading a variety of pollutants and can take up some contaminating metal ions. A different approach is to grow organisms which bring about physical changes in waste waters to make polluting metals precipitate in forms that are easier to extract.

QUESTIONS

5.22 You are provided with a fungus known to degrade some plastics used to make carrier bags, and two carrier bags. One carrier is described as biodegradable, the other is an ordinary bag. Devise a procedure to investigate the biodegradability of the two carrier bags by your organism at soil temperatures. Make a statement of the hypothesis you are testing, outline your procedure, identify the variables you need to control, and explain how you will control them.

5.23 How are microorganisms used to reduce oil spills?

'WHERE THERE'S MUCK THERE'S BRASS' – TURNING WASTE INTO PROFIT

5.9

Mining with microorganisms

As you read in the previous section, microorganisms can be used to extract metal ions from their immediate environment. **Phytomining** uses the ability of plants and microorganisms to take up mineral ions from the environment, and bioaccumulation by organisms could be a way of extracting valuable metals such as nickel or thallium from low concentrations in soil. Potentially, the energy for the extraction process could come from burning the plant material.

Bioleaching uses microorganisms to alter the physical conditions in an ore so that the desired metal ions are released from compounds in a form that can be easily extracted and purified. The process is a proven commercial technology for copper extraction from low grade ores and is now being applied to the extraction of a number of other metals such as cobalt, gold and nickel. Uranium has been extracted microbially, though it is not an economic process. The process is very similar to the copper extraction outlined below and also uses *Thiobacillus ferrooxidans*. At first there were doubts that biological processes would be robust enough for the harsh environment, but they clearly reduce the energy needed for extraction processes and leave less polluting waste at the end. Some processes can even extract useful quantities of metal from wastes generated by more conventional techniques.

Copper extraction

Copper is one of the most important industrial metals; even low grade copper ores are worth mining as the demand is so great. The ores are known as porphyrics and contain 2–5% copper. Chalcolite is the most commonly mined ore, which is copper sulphide (Cu_2S). Conventional copper production involves three processes. Extraction of the ore and smelting are done at the mine if power is available, and both use large quantities of coal. The impure metal is relatively cheap when sold on to the refiners who refine it by electrolysis. The mining process is expensive, energy consuming, polluting, dangerous and very unsightly.

The biology

Microorganisms are used to extract significant quantities of copper. The technology is old – the Romans are said to have used it – but simple. Microorganisms are used to release copper ions from ore and alter the physical environment in the ore so that copper precipitates. The process can be done at the mine, which may be in a remote or underdeveloped area, and requires very little sophisticated machinery. The organisms work effectively on low grade ores and can extract copper from ordinary mine spoil heaps. The process can even use up scrap metal.

The microorganism is *Thiobacillus ferrooxidans*, which is a chemoautotroph. It carries out inorganic oxidations to gain energy for metabolism. In doing so, it creates conditions that release copper in a soluble form. In damp conditions, *T. ferrooxidans* oxidises sulphides such as iron(II) sulphide (pyrites) to iron(III) sulphate, an energy-releasing reaction. Sulphuric acid is made at the same time, making the environment very acidic. In the copper extraction process both these products play a part in the extraction of copper from copper sulphide as copper sulphate.

The bioleaching process

The crushed ore, mainly copper sulphide with iron pyrites impurities, is heaped up on a waterproof surface. It is sprayed with water acidified with sulphuric acid to start the process, though once underway, drainage water from the dump can be recycled. The microorganisms are already present in the soil and ore at low levels and the acid conditions encourage their growth. They oxidise both the ore and iron pyrites impurities releasing iron(III) ions. The copper compound from the ore reacts chemically with iron(III) ions releasing copper. At the same time, sulphate ions are formed which react with copper to make copper sulphate, which drains out of the dump into a shallow pond. Scrap iron thrown in the pond displaces the copper from copper sulphate and copper metal precipitates out. Precipitated copper is scraped out and refined electrolytically. The process takes up a lot of space and is unattractive, but it does not require high technology, furnaces or large amounts of fuel. Figure 5.7 shows the bioleaching process. The same reactions are being used to clean up waste water draining from worked-out mines that contain high levels of metal ions.

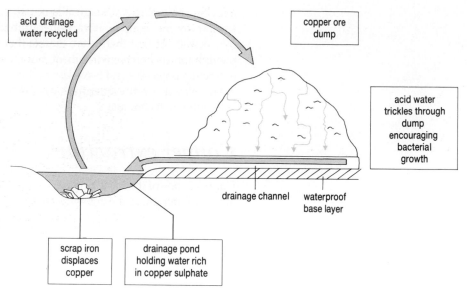

acid drainage water recycled

copper ore dump

acid water trickles through dump encouraging bacterial growth

drainage channel   waterproof base layer

scrap iron displaces copper

drainage pond holding water rich in copper sulphate

FIG 5.7 Copper is extracted from low concentrations in ore by bioleaching.

> ### QUESTION
>
> **5.24** Why can mining copper with microorganisms be a cheaper process than the conventional method?

Fuels

Energy is released when carbon-containing compounds are oxidised by burning. For most of human history we have used **biomass** as an energy source. Burning wood, dried vegetation, cow dung, cereal straw, coconut husks, oils and fats provides heat, cooking power and light. Obtaining energy in this way uses **renewable** sources – that is, the energy source can be replaced by cultivating plants and animals. In many parts of the world deforestation has led to a wood shortage and people use other fuels such as dried cow dung. Biomass fuels could provide a significant proportion of any country's energy needs if properly managed. Pilot projects using fast-growing trees for coppicing on marginal lands are being investigated for biomass fuel. However, these fuels are often bulky and awkward to use on an industrial scale. Nevertheless, biomass in the form of organic wastes can be used for power. Sewage plants already use sediments to generate methane and there are other methods of using a processing waste to make fuel. These processes are most useful in places where oil has to be imported, and places where the population is scattered in small communities far from central power generators. Major programs generate alcohol from biological materials, and methane in biogas.

Alcohol

Ethanol and other alcohols are raw materials for the manufacture of plastics and other synthetic materials. Ethanol can also substitute for mineral oils, and is used as a transport fuel, see Figure 5.8. Petrol is mixed with varying proportions of ethanol: up to 10% dehydrated ethanol is added to petrol as an octane enhancer to replace lead compounds; 20% added dehydrated ethanol makes **gasohol**, used as a fuel in its own right. Cars can also be made to run on pure ethanol. Countries that have to import oil for petrol may find it worth using ethanol to extend, or substitute for, petrol, if it can be made cheaply enough. Obviously the cost of ethanol manufacture is crucial; power for production and raw materials must be cheap.

FIG 5.8 In Brazil, gasohol is sold at garages just like ordinary petrol. Gasohol was subsidised at first to encourage people to switch to the new fuel.

Ethanol is usually made as a product of the oil industry, but alcohols can also be made by microbiological methods, using microorganisms that ferment sugars into ethanol. During World War I, for example, butanol and acetone for explosives were made by *Clostridium acetobutylicum* fermenting starches and molasses anaerobically. Microbial production of ethanol from biomass is under active development.

Many carbohydrate-rich wastes are suitable for ethanol production. They include straw and processing wastes such as maize cobs, fruit or vegetable peelings, leaves, rice straw, pods and husks. Other industries such as wood, paper and cotton processors produce pulp liquors, sawdust, and fibres that contain waste carbohydrates. Industry wastes cost money to deal with and may simply be dumped, causing gross pollution. It is economically advantageous to find a use for it.

Finding a suitable fermenting organism is a problem. Most ethanol-producing microorganisms cannot use cellulose or starches, so carbohydrates have to be hydrolysed into sugars first. Though many organisms can ferment sugars such as glucose and fructose, few can degrade the smaller 5-carbon sugars made in the breakdown of hemicellulose. Genetic modification of *Zymomonas mobilis* looks promising, and so are efforts to transfer the two genes involved from *Zymomonas* into *E. coli*, which can degrade xylose from woody materials too. There are also research studies into various ways of using microorganisms with an acid hydrolysis, or with cellulases.

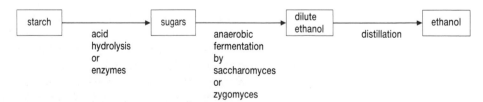

FIG 5.9 The breakdown of starch to ethanol.

Currently yeast is the most commonly used organism. Waste carbohydrates are broken down to simple sugars, or **saccharified,** by acid hydrolysis, or by enzymes. The basic fermentation process, shown in Figure 5.9, works best at 30–35 °C. At the end of the fermentation the solution contains about 9% ethanol, which is extracted by distillation. Power costs for the distillation may be the deciding factor in the economics of the process as a whole.

Gasohol in Brazil

Brazil can supply only part of its petrol needs from its own oil fields but a programme to replace petrol with alcohol has successfully reduced oil imports. Production of pure ethanol fuel for cars started in 1975 with a target of saving 40% of petrol consumption, about 10 billion litres, by 1985. As ethanol was substituted, petrol imports were held level in the face of increasing demand. At the start, the programme was only just viable, but by 1983 the world oil price rises had made the process economic. Eventually the programme led to a drop of 20% in Brazil's oil consumption. However, a subsequent drop in oil prices made it uneconomical to make and run all-alcohol cars, but gasohol remains and billions of litres of alcohol are produced each year.

Getting the programme started

Existing petrol pumps and cars had cheap, easy to carry out modifications. The new fuel was sold at half the price of pure petrol as an incentive to people to switch, and cars specifically designed to run on gasohol mixture or on ethanol were manufactured. Cars with the special engines can run on pure ethanol, but it was usually blended with at least 3% petrol to stop people drinking it! Around 2.5 million cars were running on the two fuels by the early 1990s. Ethanol fuel is less polluting to the atmosphere than petrol, producing less carbon monoxide, nitrogen oxides and sulphur oxides; neither does it contribute to photochemical smogs. In fact, the use of ethanol substantially reduced the production of greenhouse gases from Brazil.

The process

The country's under-employed sugar mills were used and distilleries were built to concentrate the ethanol. The crucial factors in the production of ethanol are the source of cheap fermentable carbohydrate and the power costs of the distillery. One of Brazil's main cash crops fulfils both these needs and solves a waste disposal problem at the same time. After harvesting, rollers crush sugar cane and the juice is extracted. This leaves the fibrous part of the cane as a waste called **bagasse**, which can be dried and a proportion is used to make animal feed, hardboard or filler. Cane juice is processed and sugar crystallises out. It is extracted two or three times, leaving a syrup containing fructose and glucose, called **molasses**. It is uneconomic to try to extract more sucrose, so the molasses is used for animal feed or other uses, or is dumped. In ethanol manufacture, pasteurised molasses is fermented by *Saccharomyces cerevisiae* to produce impure dilute ethanol. The fermented liquor is centrifuged to separate the yeast, which is recycled. Burning dried bagasse powers the distillation that produces anhydrous ethanol.

Some implications

There are always advantages and disadvantages to any process.

- When Brazil diverted its molasses into alcohol production, the world price of molasses for animal feed went up, so it became worth selling the molasses again and buying oil!

- Anhydrous ethanol could be better used as a raw material in the chemical industry to make ethylene for plastics, thus reducing the import bill for petroleum derivatives.

- It was calculated that it would need 15% of the world's production to reach Brazil's 1985 target. Sugar cane takes a long time to grow; it is a seasonal crop and can be harvested only twice. The programme needs a continuous supply of fermentable material together with cheap power for the distilleries.

- Alternative crops such as manioc (cassava) could be used but it contains starch that has to be converted to sugar before fermentation. This begs the question of whether a farmer should be eating his manioc or using it to drive his tractor!

- Problems have arisen with rainforest clearances and the displacement of small farmers from their land by larger growers. The programme would require a substantial proportion of the land currently cultivated to make Brazil self-sufficient.

- In 1989, growers' discontent over the price they received led to shortages and supply problems, which knocked consumer confidence.

5.25 What is gasohol and what is it used for?

5.26 Why don't people living in Great Britain use gasohol?

5.27 What incentive was used in Brazil to encourage people to switch?

5.28 If cars can run on pure ethanol, why is 3% petrol blended into it?

5.29 Would it be sensible for a sugar-producing country like Brazil to rely entirely on molasses to supply the ethanol program?

Biogas digesters

Biogas digesters generate a gas from organic wastes that is burnt to provide power. It is usually cheaper for industrialised countries to generate power from other sources, but using waste to generate methane solves a waste disposal problem. The process is most useful in developing countries where fuel is expensive or firewood is scarce, or there is a sewage disposal problem. The generated gas is a mixture of methane and carbon dioxide that can be used for cooking, lighting, as a substitute for diesel in generators, and for powering refrigerators. The sludge left at the end is a useful fertiliser free from pathogens or weed seeds, which are killed by the high temperatures generated in the process. It does not smell and can be stored. Dried dung burnt on a fire is only one-sixth as efficient an energy source as the same dung used in a fermenter. The process helps reduce deforestation and soil erosion, and increases soil fertility. The people who need methane the most often cannot afford to buy, run or maintain high technology fermenters, so efforts are directed to producing appropriate designs for poorer people.

The process

FIG 5.10 A domestic-scale biogas fermenter on trial in India. A range of organic waste can be fermented to produce a mixture of methane and carbon dioxide which is held in a reserve until needed.

There are many designs, from simple plastic bags laid in trenches to concrete-lined pits with input hatches and outflow vents and regulators, see Figure 5.10. The gas flow is uneven, so a storage device with a regulator valve is used to stop the biogas cooker going out half way through cooking dinner. Better designs are under investigation in which the fermenter contents are stirred or the bacteria held on a fixed film.

Organic wastes including human and animal excreta, vegetable peelings and animal wastes can go into the digester. Methane-producing bacteria are anaerobic and live in cows' digestive systems, so cow dung in the mixture provides a starter culture. The digesters need to be kept warm, at least 15 °C, and oxygen-free for fermentation to take place. Anaerobic bacteria first degrade the organic matter to methanol, hydrogen, formate and ions, then methanogenic bacteria use these to generate methane and carbon dioxide, a mixture known as **biogas**. The exact composition varies but over half of it is methane. Methanogenic bacteria include *Methanobacterium* and *Methanococcus*. Maintaining the fermentation is difficult as the methanogenic bacteria are sensitive to many substances that could enter in the refuse, such as detergents, high fatty acid concentrations, heavy metal ions and sulphides.

5.30 How can biogas contribute to the household economy of a family in Nepal?

5.31 List four materials that could be used as start materials for the production of biogas and gasohol.

5.32 Construct a flow chart to show the chemical changes that convert organic substances to biogas.

5.33 How can microorganisms be used to help the world population overcome the problem of diminishing natural resources?

QUESTIONS

Exam questions

5.34 Farm waste can be a considerable pollution problem. However, it may be a useful substrate for bacteria to ferment. In Asia, farm waste fermenters provide power for remote communities.

    (a) Name the product from such a fermentation which can be used to provide power. (1)

    (b) (i) Name a bacterium that can be used in such a fermenter. (1)

        (ii) Suggest a source, other than farm waste, for a bacterium that could be used to inoculate the fermenter. (1)

        (iii) State the conditions of oxygen concentration under which these fermenters work. (1)

    (c) Explain why these fermenters are not more widely used. (2)

    (d) Suggest a use for the residue remaining after the fermentation. (1)

OCR Sciences, March 1999, Paper 4806, Q. A1

5.35 **(a)** Describe how a **named** microorganism is able to extract copper from low grade ores. (10)

    (b) Discuss the advantages of using genetically engineered microorganisms in dealing with industrial waste and pollution. (6)

OCR Sciences, June 1999, Paper 4806, Q. B2

SUMMARY

Natural waters and soil carry microorganisms that degrade organic material arising from death and decay.

Microorganism numbers are low in natural waters but rise when organic matter contaminates the water, and different species are found.

Contaminated waters are aesthetically unacceptable, carry disease-causing organisms, and disturb aquatic ecosystems. Water purification is designed to remove particulate matter, degrade organic material, make water aesthetically acceptable, and to reduce the numbers of harmful organisms.

Water is purified by filtering, by holding in reservoirs or by treatment processes, or by a combination of treatments; then it is chlorinated for distribution.

Waste water is treated to reduce organic matter, nitrogen-containing ions, suspended particles and harmful organisms before it is returned to water courses. Organic material in effluent is degraded by the enzymic action of a variety of microorganisms. Sewage can be treated in several ways, all of which use microorganisms to degrade organic material in aerobic conditions.

Wastes including sewage sludge are used to generate fuel in the form of methane or ethanol.

Microorganisms can be used to remove polluting contaminants from soil and can be used to concentrate metals in a less polluting process than conventional mining.

6 INDUSTRIAL MICROBIOLOGY

LEARNING OBJECTIVES

After studying this chapter you should be able to:

① explain why cells and microorganisms are used to make products

② describe how microorganisms are grown on a large scale to make commercial products

③ understand some of the problems that have to be overcome to grow microorganisms industrially

④ describe processes used to separate the product from the culture

⑤ describe how specified products are made.

6.1 PUTTING CELLS TO WORK

This chapter is about commercial processes using microorganisms or their products. However, note that food and fuels made with microorganisms were covered in Chapters 4 and 5, and that plant biotechnology is also in Chapter 9. Industries using microorganisms or enzymes, such as brewing, flax fibre extraction and tanning, have a long history. Enzymes and antibiotics , agricultural chemicals, vitamins, vaccines, diagnostic agents, pharmaceuticals and flavour production by microorganisms have all become important processes. Huge opportunities for new products and more efficient processes arose when scientists discovered how to modify a microorganism's genes to make better or different proteins. One of the earliest genetically modified products on sale was a form of human hormone called Humulin offered by Eli Lilly in 1984.

Industrial processes use microorganisms in different ways. Some organisms secrete the product into their surroundings, others may carry it within their cells. Some microorganisms are used to carry out chemical conversions on other substances. The metabolic processes involved may be part of a microorganism's normal repertoire of activity, or it may have been genetically 'tweaked' to make a better version of the product, or to yield more product, or even an entirely new substance. You can read more about how bacteria are genetically modified in Section 6.5.

Why microorganisms are useful for industrial production

Manufacturers may use a microorganism instead of a chemical synthesis to make a product because

- microorganisms carry out chemical processes quickly at near-ambient temperatures. The same reactions in a laboratory may need high temperatures, pressures or a catalyst;

- some products, for example citric acid, are difficult to synthesise from raw materials but microorganisms can make them;

- microorganisms have a very rapid metabolic rate, quickly using the substrate available and turning it into product;

- microorganisms' enzymes are specific and produce fewer by-products than a chemical synthesis, simplifying purification and increasing yield;

- there is a wide range of reactions available, and the adaptability of microorganisms enables them to grow in a variety of conditions;

- some substances, particularly pharmaceuticals and hormones, can't be synthesised at all and have to be extracted from living cells;

- enzymes from living organisms make molecules with the correct orientation (left-handed and right-handed molecules), unlike industrial processes;

- there are high costs involved with chemical synthesis;

- there are fewer limits on the volume of production in a microbial synthesis, so very large quantities can be produced;

- processes using microorganisms can be cleaner technology in terms of reduced fuel costs, fewer pollutants and wastes to dispose of, less toxicity to life and less overall contribution to global warming.

6.2 GROWING CELLS ON A LARGE SCALE

If we want to use a microorganism on a large scale, we need to know a lot about its growth, its nutrient requirements and how its metabolism can be influenced to make the desired product. The conditions must be controllable to keep the organisms growing in the optimum way. The product must be exactly the same as when it is made on a small scale; the molecular shape and active sites must be the same, especially if the product is to be used for people. Neither should it have harmful effects on users or process workers. The product must also be consistent from one batch to the next. On top of all this, there should be the minimum of processing or purification, which adds to costs.

Microorganisms are grown in **fermenters,** which are tanks like those in Figure 6.1, holding from a few hundred litres up to thousands of litres of culture medium.

FIG 6.1 Industrial fermenters hold thousands of litres. The linked pipework delivers nutrients, acids, bases, and air to the culture, as well as venting gases and supplying steam for sterilising. All parts are carefully monitored.

The process is called a fermentation even though the microorganisms may be using aerobic respiration. The microorganisms are supplied with nutrients and the environmental conditions are carefully controlled to keep them growing in a state in which they will make maximum product. The fermenters are usually stainless steel because many processes produce acids, which would corrode ordinary steel, but occasionally special alloys are used if particular metal ions interfere with the product. A simplified diagram of a fermenter is shown in Figure 6.2. The pipework and valves are arranged so that parts of the system can be isolated and sterilised if it becomes contaminated, without having to shut the fermenter down and lose an entire production run.

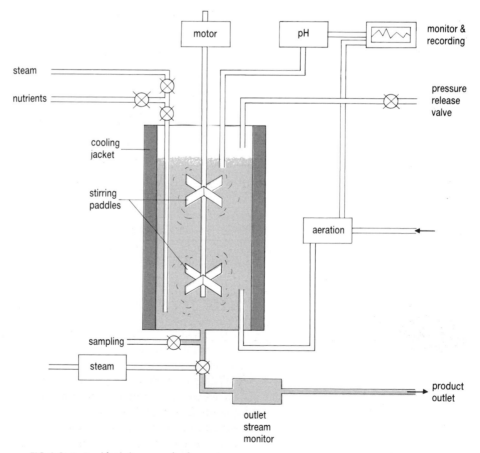

FIG 6.2 A simplified diagram of a fermenter.

Figure 6.3 shows the general manufacturing process as a flow chart, though individual products may need slightly different processing techniques. The three main stages are the fermentation; downstream processing, which extracts a pure product from the culture; and product packaging for a long shelf life.

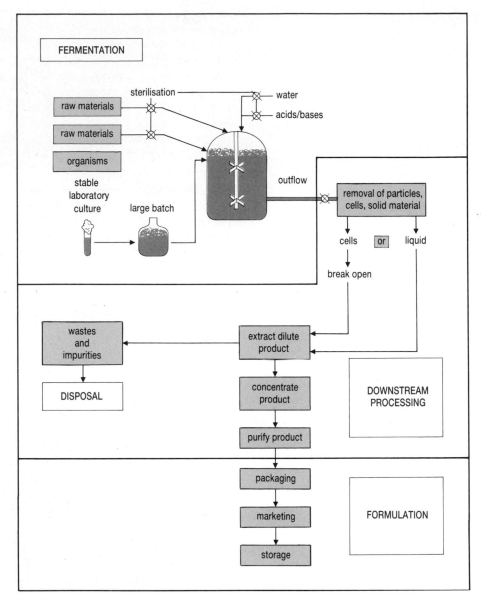

FIG 6.3 The manufacturing process.

The fermentation

The fermenter's interior and all pipework are sterilised before production begins. The usual method is to wash out and flush through with steam, which kills organisms and most spores.

Most processes are batch cultures, that is all the nutrients and a culture of microorganisms are mixed at the start of the process and kept in optimum conditions until enough product has been made, or nutrients run low, or culture conditions become unfavourable. A **starter culture** is prepared from freeze-dried stocks, or previous cultures, and grown as a small batch. The organisms are from genetically stable cultures that do not change during the fermentation or between batches. They are also robust enough to stand the conditions in a large fermenter.

Nutrients and other raw materials are sterilised then fed into the fermenter through valve-operated pipelines at rates that are monitored and regulated to control growth. Steam is blown into the fermenter for a short time at a controlled rate to warm the nutrient solution before the starter culture of microorganisms is added. Once the process is started, a **cooling system** is vital because of the heat produced by microbial respiration. An industrial scale antibiotic-producing fermentation can raise the temperature by 1 °C per hour or more. More heat comes from the stirring paddles and other mechanical movements. All this activity quickly generates temperatures of 60 °C or higher, which can kill the microorganisms and denature enzymes, bringing the process to a halt. The heat may affect protein products too. A water jacket around the fermenter, or cooling coils inside it, cool the mixture.

Aeration is needed for processes using aerobic microorganisms. Incoming air is filtered and pumped into the base of the fermenter where it bubbles up through the mixture. A valve vents gases at the top of the fermenter. Outgoing gases are filtered too, particularly if the organisms are potentially harmful. Air bubbles passing through the medium may be enough to circulate its contents, ensuring an even distribution of nutrients. Bubbles entering the mixture make it less dense locally and a current is set up as the medium is swept up with the bubbles. You can see this in the pressure cycle fermenter in Figure 6.4. Anaerobic processes are similar but the oxygen concentration may have to be reduced, or replaced with carbon dioxide if the organism is sensitive to oxygen.

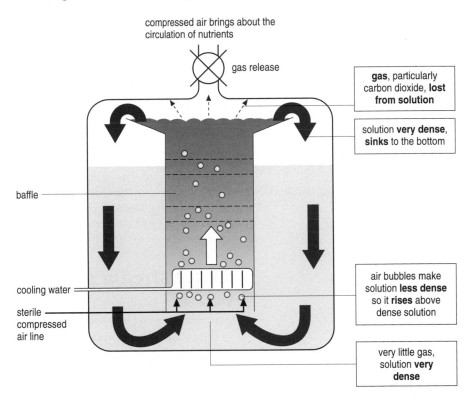

FIG 6.4 A pressure cycle fermenter uses differences in density of the culture medium to mix the contents.

Thick or sticky substrates need stirring by a motor-driven paddle called an **impeller**. **Anti-foaming agents** may be added because some nutrient mixtures make foams when stirred due to surface-active chemicals in the broth. The foam may block the air vent and any cells caught in it have few nutrients.

Once the microorganisms have reached the desired growth phase, conditions are kept as close as possible to the optimum. The conditions within the fermenter are monitored by sterilisable **probes** that record temperature, pressure, stirrer speed, pH, oxygen and carbon dioxide levels. Electronic control systems with automatic valves regulate environmental conditions; for example, if the medium becomes too acidic for optimum growth, bases are added from a reservoir to correct the pH.

Bacteria grow as single cells dispersed through the medium, but fungal hyphae spread through the medium making it viscous and difficult to stir. A mycelium breaks up into pellets if it is stirred vigorously enough, each pellet acting as a single manufacturing unit. Eventually the amount of product made declines and the fermentation is stopped. The microorganisms may still be growing well but the extra product made per gram of culture per hour becomes less economical. The fermentation is stopped, often by raising the temperature. The fermenter is shut down, emptied, cleaned and sterilised.

In a **continuous culture** system, nutrients are replaced and culture medium containing the product is drained off. Continuous culturing enables manufacturers to run their fermenters continuously, which is more economical than stopping and starting, with the equipment idle for long periods while it is cleaned and sterilised. It also ensures that there is a continuous supply of product and the rate of production can be changed according to demand. As smaller amounts are drained out at any one time, smaller equipment and fermenters are used.

Some products are made in multi-stage processes, involving activities in different fermenters. In this case, the fermenters can be linked together or the processes organised so that all batches finish at the same time, keeping the machinery in use as much as possible.

There are problems with products that do not dissolve well in water, for example steroid hormones. These have to be produced using organic solvents in substrate emulsions. Though many enzymes work in these solvents, living cells suffer because of the effects on the cell membrane. One solution to the problem that can be used with yeast is to immobilise its cells in a polyacrylamide gel, or a matrix made of alginate or collagen. Immobilised cells do not multiply much but can be used for long periods. The substrate is trickled over the matrix where microbial enzymes act on the substrate to make the product. The product is extracted from the medium as it drains away from the immobilised cells. Using microorganisms this way gives a product uncontaminated by cells so there is less purification, and the cells have a longer useful life. This is similar to the way enzymes are used, see Section 6.9.

Special precautions are taken if the process uses pathogenic organisms or genetically engineered organisms, which are subject to special containment rules. All entry and exit points, observation and sampling ports are sealed to prevent escape to the environment.

Downstream processing and packaging

The product is extracted from the culture, concentrated and purified before packaging. These processes add to the costs and may affect the nature or stability of the product. The first step is to separate the microorganisms from the medium by centrifuging. The next stages depend on whether the product is secreted into the medium, is contained within the microorganisms, or the microorganisms themselves are the product.

- If whole cells are wanted, they are separated from the medium and washed before packaging ready for their purpose.

- Microorganisms are broken open to release products from within the cells or in cell walls. This is done by mechanical shearing or by chemical treatments.

- Substances secreted into the medium are concentrated.

In the last two cases, the product has to be separated from the culture medium and any contaminating cell debris. Evaporation and crystallisation are useful techniques but products such as proteins are heat sensitive and so other techniques are needed. Water can be removed from a solution by osmosis. There are several variations on the technique but essentially a dilute solution containing product passes along one surface of a semi-permeable membrane and a concentrated solution passes along the other. Water passes from the dilute solution by osmosis, leaving a more concentrated solution of product. Hydrostatic pressure through filters is also used to remove water. Product molecules can be extracted using adsorbent columns from dilute solutions, and migration through electrical fields is useful for charged molecules. Particularly valuable products, for example interferon, which form less than 1% of the crude mixture can be extracted using monoclonal antibodies in immuno-adsorbent columns.

Packaging depends on what the product is to be used for. Vaccine materials are dried, if possible, for a longer shelf life and packaged in ampoules for dispensing. Washing powder enzymes are packaged in coloured inert pellets for mixing with powders, but are also used as liquids in many other washing products. Vitamins, flavours, sugar syrups and food additives are often used as concentrated liquids to be added to batches of food in production processes. Antibiotics are usually packaged as powders for further formulation by drug manufacturers.

 SPOTLIGHT

The problems of scaling up

There are problems in changing from laboratory scale to factory scale processes. The fermenter must be strong walled and corrosion resistant but should not leach metals that could upset microbial growth; unfortunately special alloys are very expensive. Increasing the dimensions of the trial fermenter to a large fermenter changes the surface area to volume ratio, leading to problems cooling and aerating the mix. Continuous culture processes are preferred but it is hard to keep growth optimal for months at a time and to prevent contamination. The size and mass of the fermenter may cause building problems, such as needing extra strong foundations or having to put up giant ready-assembled structures in one piece.

The choice of substrate is tricky. A solid is often cheaper, does not take up much space in the reactor and does not produce much effluent water to treat, but it is very difficult to keep the temperature down. A liquid medium may be expensive to make up but it can be stirred. Syrups such as molasses and semi-solid media take energy to warm up at the start of the reaction and need a very powerful motor to drive the impeller through them.

If extraction and packaging is difficult, or involves many stages, or needs expensive equipment, the product has to be valuable to make it worth while. Purification processes will also lead to effluent problems. The fluid part of the medium will still contain organic material that must be treated before discharge to sewers and there may also be a large cell mass to dispose of.

QUESTIONS

6.1 Why are products such as enzymes and vitamins produced microbially?

6.2 List four environmental factors to be controlled in a fermenter.

6.3 What does an impeller do? Is there an alternative to an impeller?

6.4 Compare the advantages and disadvantages of continuous and batch culture for a manufacturer.

6.5 Briefly outline the sequence of events between growing the freeze-dried organisms and a bottle of product leaving the factory gate.

6.6 Describe two things that a manufacturer can do to prevent contamination of a product.

6.7 Reread the section in Chapter 2 on *Chlorella*, then plan an outline production process to extract chlorophyll from the cells.

The genes an organism carries in its DNA are responsible for its ability to synthesise or degrade materials. These genes encode proteins including structural proteins, regulatory proteins and enzymes, which are made to meet the cell's needs. Enzymes within the cell catalyse many chemical syntheses and degradations, all of which can yield useful biochemicals and processes. Substances made by the microorganisms, and the enzymes they make them with, are all potential biotechnological products.

Precursors and the chemicals needed in metabolic reactions, such as α-ketoglutarate in respiration, are called **primary metabolites**. They are found in small quantities because they are used up as quickly as they are made. Some cell products are made at particular stages in the life of a culture. For example, penicillin is made only when the nitrogen content of the substrate falls and *Penicillium* growth has slowed. These compounds, which are not vital but have a useful role, are called **secondary metabolites** and may accumulate in large quantities. Useful secondary metabolites include quinine and codeine, capsaicin – the hot ingredient in chilli pepper, and pigments such as indigo. Figure 6.5 summarises some products made by fungi from the main respiratory pathway and shows the wide potential for exploiting this pathway.

Enzymes are useful products because they can be used to catalyse a wide range of industrial processes. Microorganisms make some enzymes continually, particularly those involved in respiration and other key biochemical processes. Other enzymes are made only when needed and may not be found in cells for much of the time, for example the enzymes needed to degrade a not very common substrate. Microorganisms secrete some enzymes into their environment, for example amylases to degrade food materials or pectinase to help invade plant tissues. These accumulate in the culture medium and can be harvested, whereas microorganisms have to be broken open to extract intracellular enzymes or metabolites.

Once the regulatory mechanisms controlling cell activity are understood, they may be manipulated to make cells produce large quantities of the compounds we want. Primary metabolites do not normally accumulate, so the cell's normal control systems have to be altered so that the metabolite is made but not used. This requires very detailed information about cell activity and a delicate touch so that other vital activities aren't badly affected by the tinkering. In this and other chapters you will find out about many different products made by microorganisms. You should look at your specifications carefully to check which products and processes you need to learn about in detail. Or see our website on www.bath-science.co.uk.

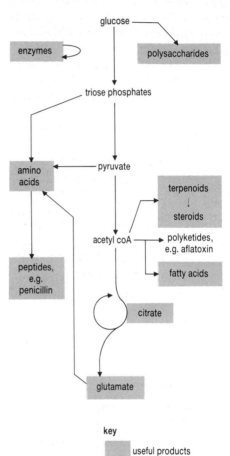

FIG 6.5 The main respiratory pathway is a source of many valuable products.

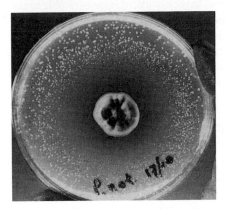

FIG 6.6 Antibiotics are secreted by fungi into the medium around them, where they inhibit the growth of other microorganisms.

We have a long tradition of using yeast, lactic acid bacteria, and *Aspergillus* fungus to make foods and so we also have a sound knowledge of how to make these organisms grow well. Similarly there is a tradition of using materials containing enzymes in leather tanning. Other uses for microorganisms have arisen as a result of scientific investigation. Bacteria were grown for vaccine production once the principles were understood. An observation by Alexander Fleming of the lack of bacterial growth around a fungal contaminant on a petri dish (see Figure 6.6) led to the development of penicillin by Florey and Chain. Once one antibiotic had been found, laboratory strains of other fungi were investigated to see if they also produced anti-microbial compounds. The search was widened to include fungi found growing wild and the antibiotic streptomycin was discovered. Most of our antibiotics have been found by checking thousands of species related to those known to produce antibiotics.

Random searches of naturally occurring species are very time-consuming and not particularly productive. Researchers can make more targeted screenings. For example, in the early development of the polymerase chain reaction (PCR, see page 103) the polymerase was denatured by the high temperatures involved in splitting open DNA molecules and had to be replaced each cycle. However, a search of the enzymes made by *Thermophilus aquaticus*, a bacterium found in hot springs, produced a heat-stable polymerase. Hot spring microorganisms produce many useful heat-stable enzymes. Sampling the soil at a site contaminated by organic chemicals turns up strains of bacteria that can metabolise them. A search in horse dung has turned up bacteria that can metabolise sewage at high temperatures.

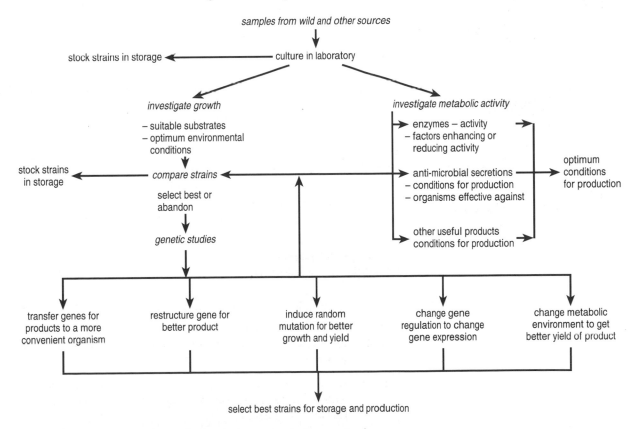

FIG 6.7 The screening procedure to develop the best strain of microorganism for a process.

Making a specialised selective medium helps in the screening process. For example, in the search for an organism to make D-proline from L-argenine, the substrate material was chemically changed to contain L-chloroargenine. It was used in the culture medium as the sole nitrogen source so that only microorganisms able to use it could grow, and they could then be further investigated.

Once a potentially useful microorganism is found, it is grown in culture, its requirements established, and the conditions needed for optimum yield of the product are determined. Various wild strains will be tried and those producing the most effective enzyme or greatest yield of product are selected and grown on. Several lines will be tested to find the lines with the greatest stability and robustness. Treatment with mutagenic chemicals can sometimes induce mutations which produce even more of the desired substance. Mammalian cell lines are treated similarly. The screening procedure is outlined in Figure 6.7.

6.5 CHANGING GENES

Often an organism's DNA is altered in some way. This may be to activate repressed genes, or to induce enzymes, or to alter the protein made by a gene to a more effective version, or even to transfer new genes into cells. For example, ethanol producers would like a fermenting yeast which has an optimal temperature of about 55 °C to complement optimal conditions for the cellulase they want to use to shorten production times and reduce the risk of contamination.

Generating mutations is a good way to obtain variations in genes but it is a random process and so the results are very 'hit and miss'. Adding the desired gene to bacterial cells from another source using recombinant DNA technology is a much more reliable technique. Bacteria that have transferred genes are described as transformed. There are many circumstances where modified microorganisms are preferred:

- when a product is very hard to source in large quantities. For example, human growth hormone is extracted from pituitary glands, which are not freely available. Factor VIII, used to treat haemophilia, is better made by microorganisms because the usual source, human blood, may carry infections;

- to make a simpler or more economical manufacturing process. Many organisms have inconvenient growth requirements or we do not know enough to grow them industrially. The desired gene is transferred into a well-understood organism grown by existing technology;

- when the usual sources are dangerous, for example pathogenic bacteria, or they potentially harbour infectious agents such as the prions thought to cause CJD. Vaccine antigens and hormones produced by genetically modified organisms will not carry any infectious agents.

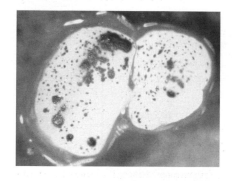

FIG 6.8 A marker gene lets the researcher know which cells have received the new genes. Sometimes resistance to an antibiotic is used; this one uses the ability to make a blue colour from a substrate compound.

Genetically modifying a microorganism involves more than transferring a single gene; most genes have controlling genes that must also be transferred. A marker gene is included too, to identify the successfully modified microorganisms. The marker gene is often one conferring resistance to an antibiotic. After gene transfer, the microorganisms are grown in a medium containing the antibiotic so that only those carrying the transferred genes with the marker gene can grow; the antibiotic kills non-transformed cells. A different marker gene produces an enzyme that generates a blue colour when the transformed cells are grown in the presence of its substrate (see Figure 6.8). There are concerns that antibiotic resistance markers could transfer through normal gene transfer processes in bacteria into 'wild' bacteria, so strict containment procedures are used for growing genetically modified organisms.

Though many microorganisms could host new genes, certain species are used more than others. These are strains of microorganisms that are safe and grow well in industrial conditions on a wide range of cheap nutrients at convenient temperatures and oxygen conditions. We also know a lot about their genome and how their gene activity is controlled. The most frequently used hosts are *E. coli* and yeast. *E. coli* is used to produce products as varied as hormones and washing powder enzymes.

'First catch your gene...'

Locating genes to transfer is difficult. A single cell, even a bacterial cell, carries enough encoded information for thousands of proteins. Finding the small stretch encoding the gene you are interested in is on a par with finding a needle in a haystack. Several approaches are available depending on what the gene is to be used for. Some methods are more appropriate for research than for generating a strain of bacteria or cells for industrial production.

Some methods start with genes in DNA. DNA carrying the target gene is extracted from cells and treated with endonucleases. These enzymes cut DNA into big sections, then **restriction endonucleases** are used to cut it into smaller fragments. Each restriction enzyme cuts between specific base sequences so all fragments produced by a particular enzyme will have the same base sequences at the cut ends. This is helpful for joining the fragments into other sections of DNA. The fragments can be transferred into bacteria, using the process outlined below, and genetic probes used to identify which bacteria carry the desired gene. This technique is laborious and generates lots of bacterial colonies for screening that do not carry the desired gene. Theoretically the libraries of gene sequences which are now available should make it slightly easier.

However, we can take advantage of the fact that genes make proteins, and the genetic code for the amino acid sequence of a particular protein is encoded in mRNA. mRNA molecules carry the information for only one protein so if you can find the right mRNA for a particular protein you have access to the coding information needed.

Specialised cells dedicated to the production of a particular protein, such as insulin-making cells in the pancreas, are rich in mRNA encoding that protein. A sample of the particular tissue is taken, the cells broken open, and mRNA extracted. There will be mRNA for other proteins as well so the various forms are separated. Once the right mRNA is identified, there are several possibilities. The mRNA can be labelled with, for example, radioactivity and used as a gene probe to match to the corresponding section in DNA fragments, and so locate the gene for transfer.

A different technique uses the mRNA molecules as templates to build up artificial DNA (called **cDNA**) with a corresponding base sequence. **Reverse transcriptase** (see retroviruses in Chapter 2), is used to make DNA from the extracted mRNA. The artificial gene isn't exactly the same as the original gene because it does not have non-coding sequences. The isolated genes can be copied by PCR techniques (see page 103) to make large quantities.

Transferring the genes

A vector is used to transfer the new gene, whether isolated from its usual DNA or made as a cDNA gene, into its host cell, but different techniques are used for transfer into bacterial, yeast, animal and plant cells.

Using plasmids to transfer into bacteria

Figure 6.9 outlines a frequently used method of transferring genes into bacteria.

Plasmids derived from naturally occurring plasmids, which are a normal means of gene exchange in bacteria, are used. Plasmids and bacterial genetic exchange are discussed in Section 2.2. Plasmids are extracted from special *E. coli* strains or other suitable bacteria. They are snipped open by a restriction endonuclease enzyme to produce DNA with 'sticky ends'. The target gene, and promoter genes for activating the inserted genes, and a marker gene are all produced using the same restriction enzyme so they all have the same 'sticky ends'. The promoter, marker and target genes are inserted into the opened plasmids. A **ligase** enzyme ensures that the fragments of DNA are joined together as **recombinant plasmid**. The 'doctored' plasmids are mixed with a culture of the host bacteria, which has been specially treated to enhance uptake of the plasmids. Plasmids enter the host bacteria in **transfection**. The host bacteria can also be described as **transgenic** if the new genes come from different sorts of organisms. DNA up to about 10 000 base pairs long can be transferred this way.

The transformed bacteria make many copies of the inserted genes – a process called **DNA cloning**. The DNA can be used for genetic research into the target gene.

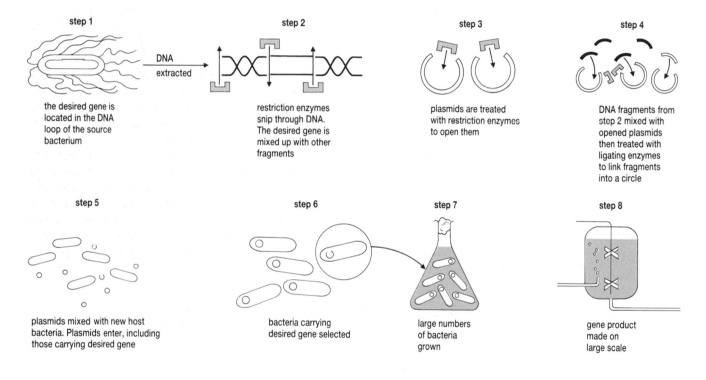

step 1 the desired gene is located in the DNA loop of the source bacterium	**step 2** restriction enzymes snip through DNA. The desired gene is mixed up with other fragments
step 3 plasmids are treated with restriction enzymes to open them	**step 4** DNA fragments from step 2 mixed with opened plasmids then treated with ligating enzymes to link fragments into a circle
step 5 plasmids mixed with new host bacteria. Plasmids enter, including those carrying desired gene	**step 6** bacteria carrying desired gene selected
step 7 large numbers of bacteria grown	**step 8** gene product made on large scale

DNA extracted

If human genes are used they are not usually snipped out of a cell. mRNA to the desired product is isolated and reverse transcriptase used to make a complementary copy of the gene which is then linked to plasmids

FIG 6.9 How plasmids are used to transfer genes into another bacterium.

Using viruses

Viruses are used to ferry genes into animal cells, and bacteriophages for transferring genes into bacteria that do not take in plasmids. In ordinary virus replication, viral DNA is incorporated into the protein coats of virus particles but occasionally other (host) DNA is also included. This process is used to genetically modify viruses to carry new genes with the viral DNA when virus particles are assembled. The new genes enter the host cell with the virus DNA when modified viruses infect their target cells. Of course, the virus has to be disabled so that it cannot cause an infection or damage the target cells. Neither should it be able to revert to a live form of virus. Adenoviruses are the most frequently used vectors. This technique has been used in trials with cystic fibrosis patients to insert correctly functioning genes into lung tissue. There is not much space inside a virus particle so only short lengths of DNA can be carried, which limits what can be transferred. For example, the dystrophin gene, a protein essential for muscle cells and lacking in Duchenne muscular dystrophy, is too large to fit into a virus. Another approach under investigation uses the desired genes encapsulated in a liposome package. The liposome merges with host cell membrane when it meets a suitable receptor and the DNA within is released into the cell.

Transfer in plant cells

There are two ways to transfer new genes into plants. In one method genes are transferred into plant cells using *Agrobacterium*, a soil bacterium which normally infects plants causing galls. It inserts its own amino acid making genes on a plasmid into plant cell DNA as it infects. Its plasmids are modified to carry new genes and gene switches, such as cauliflower mosaic virus promoter. Herbicide or Kanamycin resistance can be used as marker genes. Only cells with the marker will grow in a culture medium with herbicide, allowing the transformed cells to be identified.

A different technique uses a '**gene gun**'. Tiny beads of gold or tungsten are coated in a solution of the DNA to be transferred. The beads are fired at high velocity into the plant cells. The DNA coating the pellets diffuses into the nucleus and becomes incorporated in a tiny proportion of the cells. The gene gun technique is particularly useful because larger amounts of DNA can be transferred than with a plasmid. In some cases, genes are added that control the tissue in which the transferred gene works, for example in leaves but not seeds. There have been some successes using gene guns to transfer genes into animal cells. You can read more about why plants are genetically modified in Chapter 9.

Whichever technique is used there is still a laborious task of identifying which organisms have successfully incorporated the new material. This is done, for example, by growing in the presence of Kanamycin or another antibiotic – transformed cells will be resistant to it. Other methods use tagged antibodies raised against the particular protein made by the transferred gene, or a reaction with something in the culture medium. The successful lines of bacteria or cells are then grown in culture to make large stock samples, which are kept in storage to provide sources of starter cultures for fermentation.

 SPOTLIGHT

Making more genes

The polymerase chain reaction (PCR) gave scientists a way of copying genes rapidly. DNA encoding the gene is heated to break the hydrogen bonds holding the molecule together, and short lengths of DNA called primers are attached to each end of the section to be copied. The primers enable DNA polymerase, the enzyme that replicates DNA, to attach to the target gene. The primed target gene, free DNA subunits and DNA polymerase molecules are mixed together and the polymerase makes more DNA using the target gene as a template. The newly synthesised DNA is again heated to split it and more copies are made in a new cycle of activity. This process takes place at a higher temperature than usual in living cells, so ordinary DNA polymerase would be denatured each time the mix was heated. Instead, *Taq* DNA polymerase isolated from *Thermophilus aquaticus*, which normally lives in hot springs, is used because it is heat stable. Each cycle of DNA replication doubles the number of gene copies so that millions of copies can be made in the course of a day as an automatic process.

6.11 Outline what is meant by the following terms: recombinant DNA, plasmid, cDNA, reverse transcriptase, PCR.

6.12 Making copies of a gene to transfer may begin with extracting mRNA from cells and separating the different varieties of mRNA encoding the proteins the cell was making. Find out by independent reading how large molecules such as RNA are extracted from cells and separated according to size. Use the information you find to write a briefing paper on the process for other members of your group.

6.6 BIOTRANSFORMATIONS

Food manufacturing processes and waste management, which you met in earlier chapters, use microorganisms to carry out chemical reactions and bring about changes. In biotransformations, microorganisms convert one chemical in the feedstock into a more valuable product. There are three main uses of the process:

- something common and cheap is changed into something expensive, for example plant sterols are changed into hormones and steroids for medical use. *Nocardia* bacteria and the fungi *Aspergillus* and *Rhizobium nigricans* are used to make oestradiol and testosterone, as well as hydrocortisone derivatives from plant steroids;

- to make small structural changes in chemically synthesised molecules for a particular product. For example, peptides made by chemical synthesis are made into a form more suitable for peptide hormone synthesis, into amino acids such as proline, or vitamins; or pyrimidine-based molecules are made into precursors of insecticides;

- for products found in such tiny quantities in natural sources that it is impractical to extract large quantities.

In each case the specificity of the organism's action means that there are few by-products and purification is more straightforward. Microbial cells may be immobilised on an inert framework, which allows the same cells to work on many batches of starter material. Many organisms naturally carry the enzyme necessary for the biotransformation but other genetically modified species are being used, for example modified yeast is used in the process of making glyphosate from glycolic acid. Potentially, modified *E. coli* could be used to convert morphine and codeine into more useful painkillers.

6.7 VACCINES

Vaccines contain a suspension of microorganisms, or microbial fragments, or modified microbial toxins. You can read about how vaccines protect against infections in Chapter 7. Large numbers of microorganisms are needed to make vaccines. Early vaccines used killed microorganisms in a suspension or, as in the case of polio and rubella vaccine, a live weakened, or **attenuated**, strain. Both these types of vaccine carry a minute risk that organisms survive the process, or a weakened strain regains its ability to cause disease. Newer vaccines just use fragments of pathogenic microorganisms to provoke an immune response.

Making vaccines

FIG 6.10 Some biotechnological processes are hazardous to the workforce. This process worker making diphtheria vaccine is screened from harmful microorganisms by the protective clothing.

Bacteria for vaccines are grown in optimal conditions for antigen or toxin production. Viruses are grown in tissue culture. Rigorous precautions are taken to contain dangerous microorganisms; for example, the culture may be stirred by a magnet controlled from outside because an ordinary stirrer shaft would have to pass through an exit, which could be a source of leaks. Workers are well protected: the process worker in Figure 6.10 is enclosed within a protective suit and wears a mask. After the organisms have grown they are harvested in sterile conditions.

Bacterial cells are separated from the culture medium by centrifuging. Further processing depends on the vaccine.

■ Weakened strains are used as whole cells, which are freeze-dried to preserve them.

■ Microorganisms can be treated with formaldehyde or similar chemicals to inactivate them; detergents are used to split viruses.

■ Microbial cells such as those causing whooping cough and cholera are usually broken open to release component antigens located in cell walls.

■ The antigenic capsules of meningococci and the toxins of tetanus and diphtheria bacteria are secreted into the culture medium, from which they are extracted.

■ Toxins are treated with formaldehyde to make them harmless but capable of provoking an immune response.

Newer safer vaccines contain only antigenic fragments extracted from microorganisms. The antigen-carrying fragments are separated from the rest using immuno-adsorbent columns (see Section 6.9). There may be several different antigens, or antigens from different strains blended together to form a standard vaccine. Vaccines may provoke a better immune response when an adjuvant such as aluminium hydroxide is added. Added stabiliser gives liquid vaccines a longer shelf life. The final vaccine is checked for sterility, effectiveness, and that it is safe to use. Dried vaccines are reconstituted with sterile water before use.

Recombinant vaccines such as swine flu vaccine are made by less dangerous organisms such as *E. coli* which have been genetically modified to carry genes encoding antigens from a pathogenic microorganism. The modified bacterium makes the antigens in culture in a way that can be extracted free of cells and cell debris. There is less risk to the workforce, and recipients of the vaccine should suffer fewer side effects from the presence of other cellular materials. These **acellular vaccines** are being produced against whooping cough, typhoid, and sheep foot rot. Yet another technique uses a vaccine that contains DNA encoding viral proteins. The DNA is absorbed by cells, which produce viral proteins and stimulate an immune response. **DNA vaccines** like this are being developed against Ebola, HIV and some animal virus infections. You can find out more about the use of vaccines in Chapter 7.

QUESTIONS

6.13 Write definitions of the following terms: antigen, pathogen, recombinant. Use the index to help you find definitions in the other chapters.

6.14 Give an example of a biotransformation.

6.15 Give two advantages of genetically modified vaccines over ordinary vaccines.

6.16 Look at Figure 6.10. Explain how the process worker is protected from microbial contamination.

6.17 Draw a flow chart outlining the manufacturing process for a vaccine made with blended antigen fragments extracted from cells of several strains of a pathogenic organism.

6.8 ANTIBIOTICS AND OTHER METABOLITES

Microorganisms make useful substances as part of normal cell growth and activity. They are harvested from fermentations to make products as diverse as gibberellic acid, a plant growth hormone, and the thickener in mousse. A selection of products is listed in Table 6.1. Genetic selection and modification has produced strains of microorganisms that convert much more of the substrate into product than wild strains.

Insulin is much in demand for treating diabetes. Insulin extracted from cattle or pig pancreas is not exactly the same as human insulin and can cause side effects. The genes needed to make an insulin precursor molecule, proinsulin, were inserted into *E. coli*. The bacteria secrete proinsulin which is converted into insulin by enzyme action. The genes to make synthetic insulin have also been transferred into yeast, *Saccharomyces cerevisiae*, which is better than bacteria at making correctly folded insulin molecules, and secreting insulin into the medium. A blood clotting factor is also made using recombinant technology. Potentially, many medically useful human biochemicals could be made microbially.

TABLE 6.1 Products made by microorganisms

SUBSTANCE	ORGANISM	USE
alginate	*Azotobacter, Pseudomonas, Xanthomonas*	thickener and stabiliser in ice cream, paints, textile dressing, spray for Christmas trees, immobilising cells
avermectins	*Streptomyces avermitilis*	worming animals, removes ticks from the coat
citric acid	*Aspergillus niger*	flavouring, drinks, jam
curdlan	*Alcaligenes*	gelling agent in low-cal soups and puddings
cellulose	*Acetobacter xylinum*	cellulose sheets
dextran (1,6 glucan)	*Leuconostoc mesenteroides, Klebsiella*	medicinal use, blood plasma biochemical adsorbant in Sephadex™
dihydroacetone	*Acetobacter*	sun tanning agent
insulin	*Escherichia coli*	diabetes treatment
lactic acid	*Lactobacillus helveticus*	flavouring, confectionery, soft drinks
lysine	*Brevibacterium flavum*	enriching bread
monosodium glutamate	yeast, *Corynebacterium glutamicum*	flavour enhancer
nisin	*Lactococcus*	anti-microbial preservative
nystatin	*Nystatin*	anti-fungal drug
nucleotides	yeast	flavouring, soup, sauces
propylene oxide	*Methylococcus capsulatus*	making alkene oxides to polymerise into polypropylene and other plastics
prednisolone	*Alcaligenes eutrophus, Arthrobacter globiformis*	anti-inflammatory drug
pullulan	*Pullularia, Aureobasidium*	coating covering foods
Vitamin B2	yeast, *Eremothecium*	enrichment of human and animal food
Vitamin B12	*Streptomyces*	dietary supplement
Vitamin C	*Gluconobacter* and *Pseudomonas* in steps	food additive and dietary supplement
xanthan gum	*Xanthomonas campestris*	thickener in sauces, cheese, cosmetics, paint, ink, thickener of water for oil extraction, stabilises emulsions

Antibiotics

Antibiotics are manufactured as medicines but have other uses. For example, *Actinomyces natalensis* produces an anti-fungal agent sold under the name of Pimaricin. This is incorporated into a coating for foods such as cheeses that depend on bacterial action to develop flavour but would be spoiled by moulds growing on the surface while stored. Pimaricin affects the cell wall structure of fungi but does not affect bacteria or human consumers.

Once penicillin was discovered, other antibiotics were quickly identified. Many fungi and some bacteria, particularly the Actinomycetes, can synthesise anti-microbial agents, though some are unsuitable for use in people or animals. Many antibiotics are uneconomical to manufacture in the small quantities that may be needed, so although there are several thousand known antibiotics, 1500 from the Actinomycetes alone, only a few hundred are available commercially.

Antibiotic activity is monitored by growing bacteria in the presence of increasing concentrations of antibiotic and measuring the effect on growth; see Fig. 6.11.

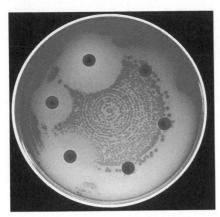

FIG 6.11 The effectiveness of the antibiotic against a test bacteria can be **assayed** by comparing the size of the cleared zone where the test bacteria (*E. coli* above) have not grown with the concentration of antibiotic.

Penicillin

Penicillin is one of the most commonly prescribed drugs in the world. It was originally isolated from a fungus, *Penicillium notatum*. It is a type of molecule called a β-lactam, which affects cell wall synthesis of Gram-positive bacteria. The original strains made very small quantities, less than 1 mg for each 2 g sugar used. *P. chrysogenum*, seen in Figure 6.12, was discovered to have a higher yield and would grow in submerged culture. Genetic selection resulted in several strains with high penicillin yield.

Penicillin is a secondary metabolite and is only made in mature cultures. Conditions in the fermenter are arranged to bring the culture to the stage where penicillin is produced, and held there. A suspension of asexual spores is inoculated into a starter medium and incubated. The starter is then added to the substrate, usually waste carbohydrate such as corn steep liquor, with additives. Oxygen concentrations are critical as the yield of penicillin is determined by oxygen availability. The proportions of various sugars in the medium affect fungal growth and can also affect the yield of antibiotic. The medium is maintained at about 24 °C and slightly alkaline for optimum growth. Penicillin is secreted into the medium once the nitrogen content of the medium falls to a low level, which is from about 40 hours onwards, when exponential growth has ceased. The fermentation is stopped when penicillin production declines. The fungus converts a large proportion of carbohydrate in the medium into penicillin, and better extraction techniques have improved the yield.

The antibiotic is extracted from filtered culture medium using solvents. The purified antibiotic crystallises. Penicillin secreted by the fungus is really a mixture of compounds, the main one is penicillin G. It can be converted to other forms of β-lactam with slightly different activities, such as ampicillin and other derivatives, and cephalosporins. Cephalosporin can be made microbially but it is cheaper to make it from penicillin. Other antibiotics are also modified after production to increase their activity or alter other properties.

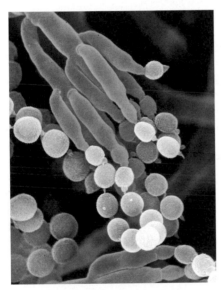

FIG 6.12 *Penicillium* species are the source of penicillin; selective breeding has changed its growth characteristics substantially for industrial production. For penicillin production the fungus grows as pellets in a submerged culture.

QUESTIONS

6.18 List two advantages and two disadvantages of using insulin from genetically modified yeast.

6.19 Refer to the section detailing bacterial cell wall structure in Chapter 2. How does penicillin affect the growth of bacteria?

6.20 Penicillin is a secondary metabolite. What does this mean?

6.21 Construct a flow chart for ampicillin production.

Microbial enzymes are used to catalyse industrial processes at low temperatures and pressures. They are useful because they are specific in the reactions they catalyse, with few by-products. For example, nitrile hydratase from *Rhodococcus* produces less than 1% by-products in Vitamin B_3 synthesis. Very small amounts of enzyme – less than 0.5% of the reaction mixture – can act on very large volumes of substrate. Enzymes are increasingly used instead of harsher processes that result in wastes with a pollution potential. By reducing the process temperature, the number of steps involved, and the amount of purification required, the costs and fuel use are reduced.

Enzymes have been in commercial use for over 100 years, originally in the textile and tanning industries, and surprisingly the first biological washing powder was formulated in 1913. Large-scale enzyme production from microorganisms is a more recent development. Enzymes are used by industries as diverse as textiles and paper manufacture, food processing, diagnostic kits, washing liquids and medicine.

Are industrial enzymes different?

Enzymes are proteins and so their activity is affected by factors that affect the three-dimensional shape of the molecule, such as changes in pH and high temperatures. However, enzymes used in technological processes must be able stand up to rough working conditions and changes in their environment. Source organisms are selected for their robust enzymes that can work in a wide range of conditions. The enzymes may be genetically modified to give them more stable structures at high temperatures and pHs differing from the norm for that enzyme. Some enzymes have exceptional properties. For example, there is a thermostable α-amylase that can degrade starches at temperatures over 100 °C. Enzyme efficiency is also improved by good design and engineering of the system. Though the exact demands will differ from one product to another, industrial enzymes should preferably

- have a long shelf life,

- work over a temperature range of 8–55 °C, with optima between 30 and 50 °C,

- have a wide pH tolerance,

- work in the presence of chemicals that usually inhibit enzyme action, such as sulphur dioxide or polyphenols,

- if necessary be able to work in a non-water solution, such as an organic solvent.

In some processes an aggregate of cells, even dead cells in some cases, is enough to give the necessary enzyme activity. More often enzymes are used as cell-free preparations, that is, extracted from the cells that produced them. Used in this way they are more efficient and the product is more uniform. Pure enzyme can be added to the substrate in solution but it is wasteful and has disadvantages. The very expensive enzymes are diluted by the substrate and are washed out of the reactor with the product when the reaction is finished. The product then needs purifying to remove traces of enzyme. It is better to use **immobilised** enzymes, which are attached to an inert support structure. The support can be porous ceramics, or the enzymes can be embedded in polymer gels, adsorbed onto membranes, or encapsulated.

The substrate circulates around and over the immobilised enzymes to give maximum contact with the enzyme. A batch of immobilised enzyme can be seen in Figure 6.13. When a reaction is finished the immobilised enzymes can be reused. Immobilised enzymes advantages are

- they are more stable; this technique mimics the way enzymes in cells are firmly attached to membranes, and only part of the molecule is exposed,
- the product is not contaminated by enzyme molecules and requires less purification,
- immobilised enzymes can be used in continuous processes,
- the enzyme has a prolonged useful life.

FIG 6.13 Enzymes work more effectively when attached to a subsurface, and can be washed and reused.

Enzyme production

The fungus or bacterium that produces the enzyme is grown in well aerated batch cultures, except for glucose isomerase by *Bacillus coagulans* which is a continuous culture. Most organisms are grown on a sterile starchy medium such as potato starch in a buffered solution. Glucose in the medium can repress production of some enzymes so it is omitted. After 2–5 days, when enzyme production and growth slows down, the temperature is raised, enough to kill the cells but not to damage the enzyme. Some enzymes are secreted into the medium from which they are extracted; others are in the cells. The intracellular enzymes are released by grinding up the cells. Particles and cell debris are removed, and the solution concentrated by evaporation or ultrafiltration. Enzymes are available as pure preparations but often an impure mixture, which is cheaper to make, is good enough. The enzymes are then formulated and packaged for distribution. Some important microbial enzymes are listed in Table 6.2. You can read about chymosin, used in cheese making, in Section 4.5.

TABLE 6.2 Some enzymes produced by microorganisms

ENZYME	USE	ORGANISM
α-amylases	break down starch to dextrins in mash, some produce maltose, degrade stains on clothes, paper manufacture, desizing starches on textiles, do not damage fibres, thicken canned sauces, improve flour by degrading starch to sugar, reduce rate of bread going stale, help separation of loaves, vegetable juice extraction, formation of glucose syrups from carbohydrate	*Aspergillus oryzae, Bacillus subtilis, Bacillus licheniformis, Actinoplanes*
glucanase	degrades glucan in beer, prevents pipe blockage and cloudiness	*A. oryzae, B. subtilis*
catalase	preservative in soft drinks, oxygen for foam rubber making	*A. niger*
cellulases	washing powder colour brightener, animal feed from rye grass and straw, stonewashing denim, releasing fruit juice from pulp	*Trichoderma spp, Penicillium, Aspergillus*
α-galactosidase	degrades sugars (raffinose, etc.) in vegetable matter to release simple sugars	*A. niger*
glucoamylase	breaks down starch and dextrins to glucose	*A. niger, Rhizopus*
glucose isomerase	converts glucose to fructose, soft drink sweetener, fillings and icings for cakes, jams	*B. coagulans, Streptomyces*
glucose oxidase	preservative in soft drinks, detects glucose in diabetics' blood	*A. niger, Gluconobacter*
lactases	convert lactose to galactose and fructose for sweetened milk drinks, reduce lactose content of milk, convert whey to sweetener foods for lactose-intolerant people	*Kluyveromyces lactis, A. oryzae*
lipases	accelerate ripening of certain cheeses, modify butter oil for baking	*A. niger, Pseudomonas, Candida, Aspergillus*

TABLE 6.2 *continued*

ENZYME	USE	ORGANISM
pectinase	clarifies wine and juices by degrading pectin, releases juice, oils and colour from fruit citrus peel, extracts for soft drinks	*A. awamori, Erwinia*
proteases	break down protein stains on clothes and food residues on dishes, degumming silk without damage, flour improver, degrades gluten in strong flour for biscuit making, alteration of milk and whey proteins, bating leather to make it pliable, removal of hair from animal hides for leather, soy bean products, fruit drinks, meat extracts and preparations, gelatin, drug assay in blood	*B. subtilis, B. lichiniformis*
DNA polymerase	multiplying small fragments of DNA	*Thermophilus aquaticus*
nitrile hydratase	production of Vitamin B3	*Rhodococcus rhodochrous*
pullulanase	manufacture of dextrose and maltose syrups for use in soft confectionery, soft ice cream	*B. pullulans, Klebsiella*
microbial rennet (and chymosin)	coagulating casein in cheese making, baby food manufacture	*K. lactis, Mucor miehei, M. pusillus, M. rouxii*
sucrase (invertase)	soft centres for sweets	*A. oryzae, Saccharomyces*
streptokinase	treatment of blood clots and bruises	*Streptomyces* spp
takadiastase	clarifies wine	*A. oryzae*

Washing powder enzymes

Biological washing powders and liquids and dishwasher powders contain enzymes that degrade proteins and carbohydrates attached to textiles and tableware; many now also contain lipases. Cellulase may be included in the formulation to degrade microfibrils formed on cotton fibres during washing and wearing. This brightens the colour of the cloth and makes it feel smooth. The enzymes work in an alkaline solution, produced by detergents, at temperatures between 10 and 90 °C. They must also work in the presence of oxidising agents and optical brighteners included with the washing product. Enzyme powders have become more important with the development of fabrics that must be washed at low temperatures, and with environmental pollution concerns. They are effective at low temperatures and work on stains that low phosphate detergents do not cope with very well.

Enzymes used in washing powder are secreted into the culture medium by the producer organisms. Once production finishes the cells are separated from the medium by centrifuging and ultrafiltration. The enzymes are precipitated from the medium using salts or a solvent. They are then encapsulated with salts, binders and other materials and coated with an inert substance for safety during washing powder manufacture. The encapsulation protects the enzyme while the powder is made and protects from the action of the detergent during storage. Powders are formulated to reduce the chance of users and process workers developing allergies to the enzymes.

Sugar-converting enzymes

Processed foods may have sugar, as cane or beet sugar, added to sweeten or to enhance flavour. There are other intensely sweet materials, which manufacturers can add in smaller amounts to a product for the same sweetness. Glucose can be made by hydrolysing waste starch with acid but it is not as sweet as sucrose and may contain undesirable by-products. Fructose is much sweeter but more expensive than sugar, unless it is made using microbial enzymes.

A cheap carbohydrate such as corn starch is heat treated to gelatinise the starch, then treated with microbial α-amylase and glucoamylase which work at high temperatures. The enzymes degrade starch to dextrins then glucose. The glucose syrup can be used in a range of foods and drinks, or it can be treated to make fructose. Dextrins are a product in their own right, used for thickening processed foods. Immobilised cells or cell fragments of *B. coagulans* containing **glucose isomerase** convert glucose to fructose. Millions of tonnes of starch are converted to sugar syrups annually.

Biosensors

Biosensors are devices used to detect small quantities of biologically important molecules in complex mixtures. They give a direct and instantaneous read out. They are used in medical monitoring and food purity monitoring. Biosensors use molecules such as enzymes or antibodies that can recognise and bind to a specific target molecule. In a biosensor that uses enzymes, a reaction between the molecule to be detected and an immobilised enzyme molecule causes a change, which is transduced into an electronic signal. The amount of target chemical in the mixture can be monitored by the number of reactions taking place. Unlike ELISA systems (see Section 6.10), many enzyme-based biosensors can be re-used for repeated monitoring, though some are designed as cheap single-use diagnostic agents.

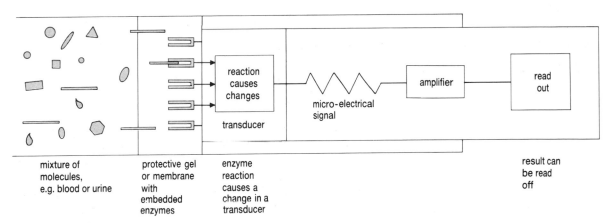

FIG 6.14 How a biosensor works.

Different sensors work in different ways. The trigger may be the appearance of a product from the reaction, or the movement of electrons during the reaction, the appearance of light or other factors. Many use oxygen electrodes but other signal transducing mechanisms have been developed. Figure 6.14 illustrates how a blood glucose biosensor for diabetics works. In the glucose-monitoring sensor, the enzyme glucose oxidase acts on glucose in the sample. Enzyme activity reduces the amount of oxygen nearby and electrodes detect the change in concentration. Other 'dipstick' biosensors are being developed for activities such as the detection of hormones, drugs, alcohols, lactate and oxygen in blood, and organic pollutants in water and soil. Biosensors are being developed that use microorganisms as the sensing device, which detect the inhibition of respiration by pollutants. Other ideas include using DNA as a gene probe to produce an effect.

Pectinase

Pectin is a large molecule (a polymer of galacturonic acid) which occurs naturally in fruits as part of the structures that hold cells together. Its long molecules make jam set, but cause gelling in fruit juices and make juice and wine cloudy. In commercial fruit juice production, two pectinases are used together, to degrade pectin in apple, cranberry and grape juice, and to clarify wine. They also help to break up fruit pulp, including citrus fruits, which increases the yield of juice from a given mass of crushed fruit. Pectinases are also used to help peel citrus fruits. Pectinases may be blended with cellulase and arabinase before being added to crushed fruit to aid juice release and haze clearing.

Lactase

Lactose is a sugar found in milk. It is an ingredient in food products, which can be broken down to glucose and galactose, again for food processing. Commercially, lactose is broken down by passing milk over immobilised lactase from *A. oryzae* or *K. lactis*.

Young mammals make lactase in the gut. Once weaned, lactase production falls in most mammals but humans continue to make it. A very rare genetic condition stops some people making lactase, and there are many others who make very little. Lactose eaten in milk products is degraded by gut bacteria instead of in digestion. The consequences of bacterial activity can cause discomfort and other problems. These conditions are called lactose intolerance. Sufferers can drink milk treated with lactase tablets to reduce the lactose content. Bacteria used to make yoghurts and cheese use the lactose for respiration, so these cause fewer problems.

SPOTLIGHT

Overtaken by technology?

Good ideas take years to come to the market, particularly with safety testing and other regulations. During the time a product is in development, new scientific discoveries can overtake it almost as soon as it is launched. *Bacillus thuringiensis* is a Gram-positive bacterium that infects butterfly and moth caterpillars. There are several strains, each infecting a different insect species. It is used as a **biological control** of insect pests in greenhouses. The bacteria carry the *Bt* gene that produces a toxic protein on a plasmid. The protein affects a caterpillar's gut cells; its blood becomes alkaline and it dies. Horticulturists can buy *B. thuringiensis* to spray in their greenhouses and on fields to reduce particular insect pests.

FIG 6.15 The sugar beet plants in the field on the left were sprayed with *Bt* toxin which kills caterpillars inhabiting the soil, whereas the plants on the right were left unsprayed. The effect on plant growth and hence yield is dramatic.

However, this is being overtaken by technology. Solutions of the isolated toxins made by the bacteria are available to spray over crop plants. The field shown in Figure 6.15 shows the impact of the toxin. The toxins are affected by sunlight so this is most useful as a short-term measure.

In a further development, crop plants such as cotton have been genetically modified to carry and express the *Bt* genes, so that insect pests are killed when they attempt to feed. There are concerns with the use of *Bt*-modified plants. One is that the widespread presence of the toxin is a spur to the evolution of a resistance mechanism; in fact, the first resistant moths were reported in 1985, well before modified plants were available.

Potentially, toxins of this type could be used to control insects that carry major tropical diseases, such as sleeping sickness, and other insect pests. Other microorganisms, including fungal spores and viruses, are also being researched as a control for insect pests.

QUESTIONS

6.22 Study Table 6.2, and list five different uses of enzymes.

6.23 What is an immobilised enzyme? Describe two ways in which an enzyme may be immobilised.

6.24 How do the properties of industrial enzymes differ from normal enzymes in living cells?

6.25 Outline a procedure you could use to determine the best temperature to wash out gravy stains with 'Whizzo' biological washing powder.

6.26 Read the section on washing powder enzymes and then construct a flow chart for the downstream processing of the enzymes.

6.10 ANIMAL CELL PRODUCTS

Cultured animal cells have been used to produce vaccines since 1947 when foot-and-mouth vaccine was made from tissue culture. Animal cells are used for basic research, to investigate the effect of drugs and other chemicals on their growth, to grow viruses, and for anti-virus vaccines. A major growth area is the production of metabolites such as collagen, tissue plasminogen activator, blood clotting factors and erythropoetin, and monoclonal antibodies. The molecular structures of these substances made by cells in culture are the same as in a whole animal so that no immune responses are provoked in the recipient.

Potentially, human cells grown in culture have many uses; replacement skin is just one example. Reconstructed human skin is a mixture of fibroblasts and collagen, incubated for a few days to form a mesh. Then a small sample of human epidermis is put in the culture. The human epidermis grows, producing a tissue that lacks many of the structures in real skin but is a very useful research tool and may one day be used for skin grafts, burns treatment and repair of damaged corneas. Research into the ability of embryonic stem cells to develop into a range of tissue types raises the possibility of generating whole tissues in the laboratory for replacement tissues and for medical research.

Animal cells are difficult to grow in a fermenter in large quantities because they prefer to be attached to something and grow very slowly in suspension culture. Immobilising cells is more successful. One method uses cells trapped in alginate beads; another uses collagen-based beads which are more porous, allowing freer movement of nutrients and products from the cells. The fermentation is still relatively small scale; reactors hold less than 100 litres or so. The animal cell growth pattern is similar to that of bacteria but large numbers of cells are needed to start the culture, and cell doubling times of around 15 hours are slow.

Embryonic tissues have the best capacity to multiply in cell cultures and have been used to develop lines of cells for laboratory culture. Stem cells are embryonic cells which still have the ability to develop into different kinds of specialised cells. Many human health problems are the result of failing tissues and organs. Potentially cultured stem cells could be used to regenerate the failing tissue.

Monoclonal antibodies

Antibodies are proteins made by white blood cells called B-lymphocytes. They are used to diagnose and treat infections, locate or identify molecules, track cancer cells and investigate foods. They are useful because they will only bind to a specific target molecule (you can learn more about these cells in Chapter 7). Antibodies can be extracted from blood serum, but only in small quantities and as a mixture of different sorts. B-lymphocyte cells and other cells of the immune system, like spleen cells, do not grow well in culture. But special lines of cells to make a specific antibody and grow in large-scale culture have been developed. It was known that animal cells in culture occasionally fuse together, particularly if certain viruses are present. A chemical, polyethylene glycol (PEG), has the same effect. The fused cells do not even have to come from the same species to make a hybrid. In 1975 Kohler and Milstein made hybrid cells by fusing mouse spleen cells with cells from a tumour called a myeloma. The cell line was called a **hybridoma**. The cell line had the spleen cell ability to make antibodies; the ability to grow in culture came from the tumour cell. This discovery enabled new cell lines to be developed which make only one sort of antibody.

The antibody-making cells come from mice that have been injected with a particular antigen so that some of their spleen cells are dedicated to making antibody to that antigen. These cells are mixed with mouse tumour cells and treated so they fuse. Only a very small proportion of the fused cells make the right sort of antibody. These are cultured on a large scale on soft agar, where they multiply to form a clone of cells synthesising antibody, hence the term **monoclonal antibody**. The hybridoma cells can be deep-frozen until they are needed. The whole process is more difficult if specific human antibody is needed because human cells have to be fused.

To manufacture antibody, suitable hybrid cells are immobilised and cultured in serum-free culture medium. Monoclonal antibody is secreted into the medium. After extraction the purified monoclonal antibody molecules can be linked to resin beads, enzymes or dye molecules and are stable enough to be used in products for the commercial market.

Diagnostic testing

Monoclonal antibodies are used in diagnostic testing kits to detect low concentrations of target. The kits are single-use, cheap compared to other diagnostic procedures and give results very quickly. They are used to diagnose infections, confirm pregnancy where they detect human chorionic gonadotrophin released by the embryonic tissues, test the purity of products as diverse as foodstuffs and medicines, and to detect drugs in the blood of sports people. In a diagnostic kit two different antibodies are generally used. One set of antibodies matches the target molecule and will bind to any in the test solution. However, the antibody–target complexes are too small to see and the second set of antibodies is used as part of the detection system. These antibodies have also been raised against the target molecule, but at a different antigen, and they are linked to other molecules, for example the enzyme horseradish peroxidase, that can bring about a colour reaction. When these antibodies meet with their specific antigen, even at very low concentrations, a colour change occurs. An alternative uses a reaction that fluoresces under ultraviolet light.

Antibody–enzyme detection systems are known as **ELISA,** which is Enzyme Linked Immunosorbant Assay. Monoclonal antibodies to the substance being monitored are attached to an inert material, a dipstick, or are in a well in a plastic test dish. They are immersed in the test solution where the antibodies form their complexes with the target molecule and the test material is washed out. Enzyme linked to a second different antibody is added to the mixture. The second antibody sticks to the target molecule complex too. Next, the enzyme substrate is added to the mixture and any bound enzyme reacts with its substrate to make a coloured product. The user sees a colour change; the intensity of colour indicates how much of the target molecule is present.

QUESTIONS

6.27 What is meant by the term 'monoclonal antibody'?

6.28 What reasons could there be for worry about the safety of products made from animal or human cells that are used in humans?

6.29 Read the section on monoclonal antibodies and then draw a flow chart for using the detection of human chorionic gonadotrophin in a pregnant woman's urine with monoclonal antibody as an example.

6.30 Monoclonal antibodies attached to radioactive materials can be used to locate cancerous cells in the body. How would you locate these antibodies when they are attached to their target cell?

6.11 PLANT CELL PRODUCTS

A quarter of our medicines and drugs are derived from plants. The top four plant drugs are steroids for the contraceptive pill from yams, codeine from the poppy, and atropine and hyoscyamine from deadly nightshade for nerve conditions. We also extract billions of pounds worth of flavours, perfumes, dyes, latex and gums, essential oils and waxes such as jojoba, and agricultural chemicals such as pyrethrins each year. Producing these compounds in controlled conditions in culture would ensure year-round supplies of a high quality product instead of irregular amounts of variable quality because of weather problems, plant pests, or political conditions in the area they grow. Useful plants may even be endangered by over collection or deforestation.

Few products are in commercial production but many areas such as the production of hyoscyamine and biotransformations are being explored. Much research focuses on investigating the biosynthetic pathways, then isolating the necessary enzymes and using them to work on precursor molecules in a chemical synthesis. In contrast, the production of plants by tissue culture techniques and micropropagation is a major industry. These are covered in Chapter 9.

Plant cells do not grow well in standard industrial fermenters for a variety of reasons, even though sucrose in the medium does reduce the need for light for photosynthesis. Glass or plastic vessels containing suspensions of cells are used, stacked closely together in racks with subdued artificial lighting. The culture medium is entirely synthetic and complex; the cells need sucrose, plant growth regulators, amino acids and vitamins. A large sample of a callus culture (see Section 9.10) is inoculated into the medium and the cells are dispersed by shaking with cellulases. The culture has to be aerated which also helps to break up large clumps. The cells follow the same growth pattern as bacteria but much slower. Very few generations pass before the culture enters the stationary phase. Single-celled algae are grown in columnar bioreactors with some success. Once again, immobilising cells seems to improve their growth. Isolated plant cells grow very slowly and tend to store the product, which has to be extracted by breaking cells open. This gives a very short useful life and makes the process uneconomic. It may be possible to put adsorption units in the reactor vessel to compete with the plant cell as a storage site for products.

6.12 TRANSGENIC ANIMALS

T he first animals carrying transferred genes were available in the 1980s. Transgenic animals are mainly used for basic genetic and medical research. The earliest genetically modified mouse carried a rat growth hormone gene. Since then, many strains of mice have been bred with genetic modifications. They are suitable for investigating specific biochemical processes, or discovering how normal genes work by using 'knockout' mice which have genes missing, or as 'onco-mice' for investigating what cellular processes are going wrong in cancers.

Genes are transferred into animals at the egg stage in order for the new genetic material to pass into cells in tissues. DNA carrying the desired genes is transferred into fertilised eggs using a micropipette. The eggs are then implanted into a surrogate mother animal and develop in the same way as other offspring.

Why would industrial manufacturers want transgenic animals? Manufacturers using biotechnological processes prefer to use familiar organisms such as yeast, which have been cultured for thousands of years, because we know a lot about how

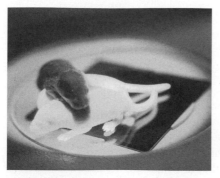

FIG 6.16 This mouse received new genes via a virus vector. It has the same problem as human muscular dystrophy sufferers. It is used to find out the genetic basis of the disease and potentially to develop a gene therapy for this incurable disease.

to keep them happy. Developing and maintaining a sterile microbiological process with a new organism to make a valuable product is lengthy, costly, and requires a skilled workforce. Keeping a flock of chickens that secretes a valuable product into their eggs is well within everyday skills and technology. Currently the processes are at the research stage with female sheep whose milk contains alpha-1-antitrypsin, used to treat emphysema. Similarly attempts have been made to genetically modify chickens to secrete important pharmaceuticals into their eggs. There are enormous difficulties in getting the gene to be active in the right tissue and nowhere else, and in some cases the gene is passed on to offspring. Other research has approached modifying animals so that their tissues could be more compatible with humans for transplant uses. There are wider social issues to be considered – would there be problems for anyone who eats a modified animal? – would the animal's welfare be affected? – do the animals have any rights to be considered?

Gene therapy

Many medical conditions can be linked to malfunctioning genes; for example, in cystic fibrosis the gene for a particular mechanism for transporting a substance out of cells (a regulator protein) does not work, with widespread consequences in the body organs. Research is being done on transferring correctly functioning genes into individuals suffering from these conditions. As the transferred genes are human, they are not actually transgenic. The delivery of genes using modified adenoviruses has been clinically trialled for cystic fibrosis, and the use of a gene gun is being investigated for a form of kidney tumour. An alternative under investigation is to make a package of lipid molecules like those found in membranes to carry the genes in. The lipids in the package would merge with the cell membrane of lung cells and deliver the contents into the cell. Under current ethical guidelines no genes can be transferred that could find their way into the eggs or sperm of the individual concerned and hence into their descendants. This rules out gene transfer at the fertilised egg stages.

QUESTIONS

6.31 What are the advantages of using immobilised plant cells to make flavours and pharmaceuticals?

6.32 Why can't plant cells be grown in fermenters like bacteria can?

6.33 Farm animals are not subject to such stringent controls as laboratory animals. It has been suggested that 'knockout' animals that have been engineered to perceive less and remember less of their environment could be developed so that they suffer less in cages and intensive farming. Prepare two arguments for, and two arguments against, developing such farm animals. What ethical issues are involved?

Exam questions

6.34 Thermophiles are organisms that live and reproduce in very hot conditions, up to 105 °C. Some are bacteria that live in volcanic vents in the sea floor. There is a great deal of interest in producing enzymes from these bacteria commercially. Samples of bacteria taken from a volcanic vent are treated to extract their DNA. Recombinant DNA technology is used to transfer pieces of this DNA to host bacterial cells, which are then screened for the production of new enzymes. The host cells are a species of bacteria often grown in fermenters.
   (a) Describe how DNA extracted from a thermophile may be inserted into the DNA of a host bacterium. (4)
   (b) Suggest and explain **two** reasons why it might be easier to use host bacterial cells to produce the enzymes commercially, rather than obtaining them directly from the thermophilic bacteria. (4)
   (c) Suggest possible advantages and disadvantages of using bacterial enzymes from thermophiles commercially. (4)

AQA Biology, March 1999, Paper BY06, Q. 8

6.35 Citric acid can be produced by fermentation, using the filamentous fungus *Aspergillus niger*. The fermentation mixture consists of a sugar (sucrose), other nutrients, and an anti-foaming agent. Large volumes of air are blown through the fermenter. After about 10 days most of the sucrose has been used up and the fermenter is emptied.

  (a) Describe **three** eukaryotic features that you would expect the fungus to show. (3)

  (b) (i) Suggest why the anti-foaming agent is required.

   (ii) Suggest **two** nutrients, other than sucrose, that would need to be present in the fermentation mixture. (3)

  (c) The citric acid is separated from the other products of the fermentation during downstream processing. Suggest **two** methods that might be used. (2)

AQA Biology, March 1999, Paper BY06, Q. 2

6.36 The diagram shows the design of one type of fermenter used to grow fungus to make mycoprotein by a continuous fermentation process.

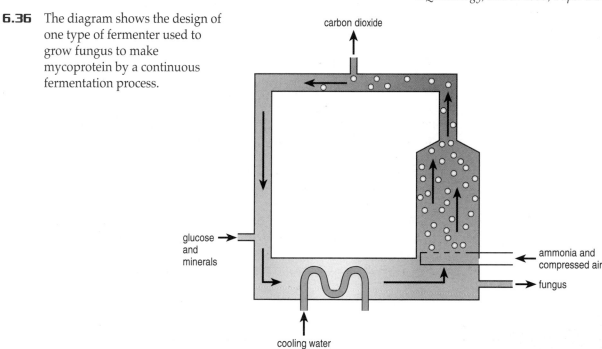

  (a) Explain why each of the following is added to the fermenter: ammonia; compressed air. (2)

  (b) Surplus heat produced during the fermentation is removed by the cooling water. By what process is this surplus heat produced. (1)

  (c) (i) Suggest how the fermenter is used to obtain a continuous supply of fungus. (2)

   (ii) Give **two** advantages of continuous fermentation, compared with batch fermentation. (2)

AQA Biology, June 1999, Paper BY06, Q. 5

6.37 Lactose intolerance in humans is the inability to hydrolyse lactose due to the lack of the enzyme lactase in the alimentary canal. The condition is inherited. The fungus *Aspergillus oryzae* is now used to produce lactase on an industrial scale. The enzyme is extracted, purified and immobilised, and then used to catalyse the hydrolysis of lactose in milk into simple sugars that can be metabolised by lactose-intolerant people.

  (a) (i) Describe how the immobilised lactase might be used to catalyse the hydrolysis of lactose in milk in a school laboratory. (3)

   (ii) Suggest how the lactase enzyme might be immobilised. (2)

  (b) (i) List **three** advantages of using immobilised enzymes. (3)

   (ii) State **three** advantages of using purified enzymes in an industrial process in preference to whole cells. (3)

   (iii) Suggest **one** reason why it may be necessary in some processes to use whole cells rather than an isolated enzyme. (1)

  (c) Lactase-treated milk is sweeter than untreated milk. Suggest why this might be an advantage in the manufacture of flavoured milk drinks. (1)

6.37 *continued*

The diagram shows the structure of lactose.

(d) Describe the hydrolysis of lactose, naming the bond which is broken and one of the products formed.

OCR Sciences, March 1999, Paper 4806, Q. 4

SUMMARY

Microorganisms and cells are used to make products because they can make a wide range of substances and can be grown in large quantities, relatively cheaply, and in a controllable way.

Microorganisms from a variety of sources are screened for potential products. Genetic modification techniques enable genes to be transferred into well-understood microorganisms more suitable for the manufacturing process.

Microorganisms are grown on a large scale in fermenters using cheap materials as nutrients. Fermenters are designed to give the inhabitants the optimum growth conditions for the synthesis of the product.

Changing from laboratory scale to factory scale introduces new problems.

Several techniques are used to extract, concentrate, purify and package the products, which may be heat-sensitive.

Products made by microorganisms include enzymes, primary metabolites and secondary metabolites.

Products may use whole microbial cells or metabolites made by cells that are extracted and processed.

Enzymes are more efficient and have a longer working life if they are immobilised on inert supports. The product is free of potentially troublesome contaminants.

Animal cells are used to make monoclonal antibodies.

MICROORGANISMS AND DISEASE

"I think great things are coming to pass. Joseph Meister has just left the laboratory. The last three inoculations have left some pink marks under the skin, . . . we approach the final inoculation, which will take place on Thursday, July 16th. . . The lad is very well this morning, and has slept well, though slightly restless; he has a good appetite and no feverishness. . . Perhaps one of the great medical facts of the century is going to take place; you would regret not having seen it."

Louis Pasteur, writing about the first human test of his anti-rabies vaccine, used to treat a small boy who had been bitten by a rabid dog, 1885.

Discoveries about the nature of infectious diseases in plants and animals, and how plants and animals protect themselves against pathogens, has led to the development of therapeutic treatments which can be used to treat infections. Other techniques such as spraying crops with protective chemicals, vaccinating people and improved hygiene have reduced the incidence of infectious disease.

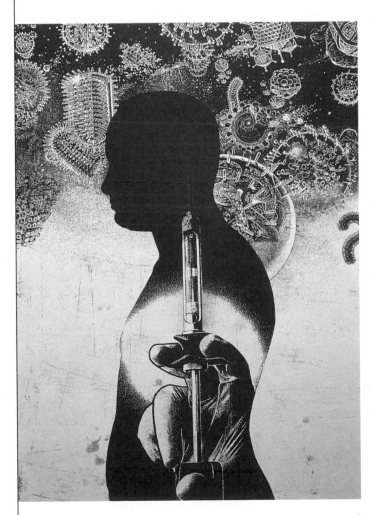

PREREQUISITES

Read Sections 2.2, 2.4–2.6, which describe the biology of disease-causing microorganisms.

7 THE BODY'S FIGHT AGAINST DISEASE

LEARNING OBJECTIVES

After studying this chapter you should be able to:

① understand how microorganisms cause disease

② explain the natural defence mechanisms in humans

③ describe how the immune system responds to infection

④ describe how defence mechanisms can be enhanced by immunisation

⑤ appreciate how drugs and antibiotics can be used to treat and control disease.

HOW MICROORGANISMS CAUSE DISEASE

7.1

Animals have co-existed with microorganisms for millions of years and have evolved defence mechanisms to stop harmful microorganisms, or **pathogens**, establishing an infection. If an infection starts, another set of defence mechanisms – the immune system – swings into action. The discovery of drugs and antibiotics has reduced the risks of bacterial infections in countries with good health care. There are fewer drugs effective against viral diseases so preventing infections is important. Malaria and other protozoal diseases cause huge numbers of infections in developing countries, and it is hard to control the spread of such diseases. You can learn about disease transmission and individual diseases in Chapter 8.

Pathogenic microorganisms are parasitic, and have a number of harmful effects.

- Bacteria and viruses may destroy tissues, with a consequent loss of function. For example, HIV infects immune system cells responsible for priming the response to infections, so sufferers cannot fight off other infections. *Salmonella* damages the gut lining and affects absorption in the alimentary canal.

- Some bacteria and fungi release toxins that affect the tissues. For example, tetanus toxin affects nerve-cell junctions and keeps muscles in spasm.

- Viruses interfere with and override normal cell functions.

- Some pathogens interfere with protein synthesis in cells or, like cholera, with transport across cell membranes.

Toxins

Toxins are soluble substances made by certain bacteria and fungi that have harmful effects on host cells at very low concentrations. Even uninfected tissues are affected because toxins are transported through the host's tissues from pathogens lodged elsewhere. Microbial toxins include some of the most potent poisons known, including botulin toxin sometimes used as a muscle relaxant (botox). **Exotoxins** are secreted by the organisms into their immediate environment. The tetanus exotoxin made by *Clostridium tetani* binds to nerve endings and blocks impulses so that the muscles of the body stay contracted in a severe spasm called tetany. The old name for the disease was lockjaw, which vividly describes its effects on the facial muscles and is shown in Figure 7.1. The enterotoxin made by cholera organisms is an exotoxin active in the gut. It affects the cells of the gut wall so they pump out valuable electrolytes and fluids, causing dehydration. **Endotoxins** are components of the outer cell structures of Gram-negative bacteria. Endotoxins also cause fever and fluid loss. Fungi make toxins too; aflatoxin, produced by *Aspergillus flavus*, which grows on stored nuts, binds with host cell DNA, blocking RNA synthesis and hence protein synthesis.

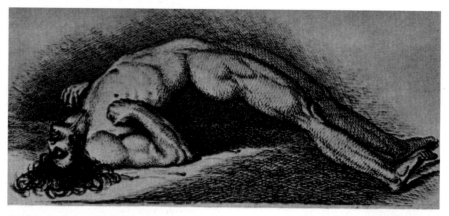

FIG 7.1 A contemporary drawing of a soldier in the Napoleonic Wars suffering from tetanus. The organism produces a toxin that affects nerves and muscles, causing muscles to stay contracted – the spasm is called tetany. Even respiratory muscles are affected.

Can the immune system become infected?

There are pathogens that are adapted to outwit the cells of the immune system, reducing the host's ability to overcome the disease. Certain bacterial toxins affect the white cells that normally ingest invading microorganisms. They either kill them outright or inhibit them so that they do not take up the pathogen. Others cause blood clotting around the infection, which bars the migration of white cells to the site of infection. TB and brucellosis bacteria can infect the white cells of the immune system and reduce our ability to overcome infections. They resist the normal processes by which white cells break down ingested microbes. HIV is a pathogen with particularly severe effects on the immune system; the virus infects the type of white cell that 'switches on' the rest of the cells of the immune system. You can read more about TB and HIV in Chapter 8.

People also suffer as a consequence of their defence mechanisms acting against pathogens. Fever and inflammation are innate responses to infection. In some infections, such as leptospirosis, the immune complexes between antibodies and microorganisms are big enough to lodge in small capillaries in the liver and kidneys, blocking the passage of blood to tissues. Some microorganisms, particularly viruses, have chemical markers (antigens, see Section 7.3) that protrude above the surface of cells they have infected. These provoke immune system cells to react against them. Hay fever, where pollen markers protrude through the membranes of lung cells that have taken them in, is one example. Another involves leprosy bacilli, which grow very slowly in human nerves. Their walls are extremely tough so components remain in nerve cells even after the organism is dead. When the body's defence mechanisms detect these components, white cells respond and damage the nerve cells in the process.

QUESTIONS

7.1 Briefly describe two ways that microorganisms can cause damage in animals.

7.2 What is a toxin?

7.3 Describe the action of one toxin.

7.4 Why do nerve cells suffer damage in a leprosy infection?

7.2 DEFENCE MECHANISMS

We have external defence mechanisms, which make it difficult for pathogens to gain entry, backed by internal defences that operate against any invading microorganisms. We also have specific defences that are activated and respond to individual sorts of pathogens. Natural entry points such as eyes, sweat and sebaceous glands, and entrances and exits to organs, are all protected by defences.

Commensals

Surprisingly, one of the best protections against pathogens is the presence of other microorganisms. Any part of the body in contact with the outside world, including the skin, gut, mucous membranes and the respiratory tract, carry billions of microorganisms called **commensals**, which are harmless as long as they are confined to these surfaces. Commensals are adapted for life around the human body and colonise the available ecological niches. Incoming pathogens, which are usually better adapted for life inside the host than outside, have to compete with commensals to become established. Commensals are kept in check by the host's normal defence mechanisms. If these fail, for example the surface tissues are damaged, or the immune system is suppressed, the commensals may cause **opportunist infections**. Transplant patients and AIDS sufferers have to be alert to opportunist infections by skin commensals.

The skin

The skin is an arid environment and a tough barrier to penetrate. The outer surface is an unstable layer of hardened cells that continuously flake off. Most of the microorganisms on the skin cluster around the sweat and sebaceous glands, shown in Figure 7.2, where there are nutrients and moisture. Anti-microbial substances such as fatty acids in the secretions inhibit microbial growth. Microorganisms can enter through wounds but clotting blood quickly makes a barrier, and white cells are drawn to the wound to scavenge infecting organisms.

Entry points

The eye surface and mucous membranes such as the nasal lining are ideal for microbial growth, as they are moist, warm, rich in nutrients and not hardened. They are protected by a variety of mechanisms, shown in Table 7.1, against pathogens in air, food and water. These are very effective; for example, most microorganisms in food and drink are killed by stomach acid but people taking drugs to lower stomach acidity suffer more diarrhoeal infections than usual because fewer organisms are killed. In general, the large intestine is more hospitable than the rest of the gut, but pathogens find it difficult to compete with the large numbers of anaerobic or microaerophilic commensals adapted to the low oxygen concentrations there.

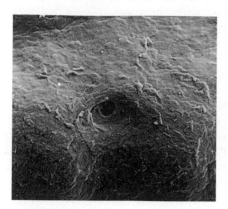

FIG 7.2 Microorganisms on the skin around the sweat and sebaceous glands.

TABLE 7.1 Non-specific diseases

AREA	PROTECTION	
skin surface	keratin	hardens cells
	fatty acids and lactic acid	inhibit microbial growth
	IgA antibody in tears, saliva and other secretions	binds to microorganisms
mucous membranes	lysozyme	degrades Gram-negative bacteria walls
respiratory tract	mucus	traps particulate matter; cilia sweep mucus up to throat so it passes to stomach
stomach	hydrochloric acid	kills pathogens
	proteases	degrade microorganisms
small intestine	bile and proteases	degrade protein and emulsify lipid components
blood	transferrin	iron is sequestrated and levels reduced to below those needed for bacterial growth

Inflammation

Inflammation is a protective response to any damage, not just damage caused by infecting microorganisms. The sequence of events brings together many different factors that localise the damage, remove damaged cells and hinder any microbial infection. The inflamed area is redder than usual, tender, swollen, and warmer to touch. The symptoms are caused by more blood flowing to the area, and the release of fluid from the bloodstream to the tissues. Capillary walls in the damaged area become more permeable and more materials can leave the bloodstream. Kinins are converted to bradykinin, which dilates the capillaries and keeps them permeable. Histamine, prostaglandins and other chemicals are released in the damaged tissues which activate white cells in that area. White blood cells can migrate through the capillary walls into the damaged tissues, drawn by the release of chemicals from a complex of plasma proteins called **complement**.

Interferons

Interferons belong to a diverse group of glycoproteins known as cytokines that activate or inhibit other cells in a hormone-like manner. Interferons are made during virus infections by white cells and virus-infected cells. They are effective against most viruses, working while antibodies are secreted. Interferon released into tissue fluid prevents uninfected cells becoming infected. It binds to the membranes of uninfected cells, stimulating them to make enzymes that degrade virus RNA and so blocking virus multiplication. Interferons also activate other cells in the immune system during the virus infection. Interferons are responsible for many of the unpleasant effects of a virus infection such as 'flu. Aches, shivers, tiredness and fevers are the typical side effects. Interferon is hard to use as a therapeutic against viruses because it is most effective when given at the same time as the virus is initiating the infection. Interferons made by cultured cells are used to treat skin cancers, multiple sclerosis and hepatitis.

Other defences

Microorganisms need iron ions to grow but find few in blood because iron is transported bound tightly to a protein called transferrin, and microorganisms cannot easily access it.

Fever, which is part of the inflammation and interferon response, is a defence mechanism. Though the elevated temperatures of a fever are potentially dangerous to people, they will kill microbial cells first.

QUESTIONS

7.5 List the main entry points into a human body for microorganisms.

7.6 Describe how the skin acts as a barrier to the entry of pathogens.

7.7 Explain how the presence of microorganisms on the skin can be beneficial in the fight against disease.

7.8 (a) Give two ways in which each of the following prevents infection: skin, tears, lining of the respiratory tract.
(b) Draw an outline of the human body on a sheet of paper. Label all the entry points. By each, note how the entrance is protected. After working through Chapter 8, add notes on one organism that infects through each entrance.

7.9 Why does increased permeability of capillary walls near the site of the infection help overcome infections?

7.10 Name two compounds that draw white cells to the site of an infection.

7.11 Give one advantage and one disadvantage of using interferon to treat a virus infection.

7.12 Self study: use a standard biology textbook to research and make a flow chart of the processes involved in blood clotting.

We have long known that people who survive a major infectious disease are usually protected against further infections. A foreign organism entering the body, or its products, generates an **immune response** and memory from the immune system. The human immune system is complex, and the roles and interactions of its many parts are still being elucidated.

The main components of the immune system are

- cells: there are several types of white blood cells, including T- and B-lymphocytes, macrophages and other phagocytic cells, and mast cells,

- tissues and organs: including the lymph nodes, spleen and other immunologically active tissues,

- molecules: antibodies and other active molecules.

How infecting microorganisms are recognised

Invading microorganisms carry proteins, lipopolysaccharides and other molecules in their outer layers. If these chemicals do not normally occur in someone's body, they are recognised by the immune system as foreign, and an immune response is provoked. The toxins and waste metabolites produced by pathogens can also provoke the immune system. These chemical markers which provoke the immune system are called **antigens**.

How does the body 'know' which are normal antigens on its own cells? Everyone carries hundreds of proteins on the surface of their cells. These vary subtly from person to person, and the particular combination someone carries is almost unique. These proteins are called **MHC** (major histocompatibility complex), and they are recognised as 'self' by the immune system. In the first few weeks of life the immune system identifies and learns to recognise these proteins and other antigens, and to ignore them. During this period maternal antibodies in breast milk and those passed across the placenta protect a baby against infection. By 2 months old, a baby's immune system is mature enough to be making its own defences, and new antigens are regarded as 'foreign' and will provoke a response. The immune system is sensitive enough to pick up quite small changes so, for example, a foreign protein belonging to a virus poking through a cell membrane beside the normal proteins will trigger an immune response. So will unusual combinations of antigens such as those on transplanted tissues or cancer cells.

Antibodies

People recovering from an infectious disease carry much higher levels of **antibodies** in their blood than uninfected people do. They also have larger numbers of **B-lymphocytes**, or B-cells, which are white cells in the blood. Antibodies are proteins made by B-cells in response to foreign material and microorganisms entering the body. They are properly called immunoglobulins; after they are released into the blood they are known as antibodies. Antibodies will bind to antigens on the surface of foreign material where they act as a marker for immune system cells, or hinder the invading microorganisms' attempts to colonise. They bind to antigens in the surface layers of bacteria, with virus capsid proteins, with membrane antigens in protozoa, and with toxin molecules.

Antibodies affect microorganisms in three ways:

- they coat and clump microorganisms together, called **opsonisation**,
- they link to receptors on phagocytic cells (see below) and stimulate them to ingest the microorganism,
- they bind to toxin molecules so that the toxins cannot bind to host cells.

Each antibody molecule matches and binds to one particular antigen only. The ability of an antibody molecule to combine with only one antigen is called its **specificity**. Figure 7.3 shows how specificity works. Each B-lymphocyte secretes only one type of antibody matching only one particular antigen; it is **dedicated** to that antigen. Antibodies released during an infection can last for months, long after the infection is beaten. A memory is retained within the immune system, so whenever more microorganisms of the same type re-enter the body, even years later, a huge amount of antibody specific to that microorganism is made.

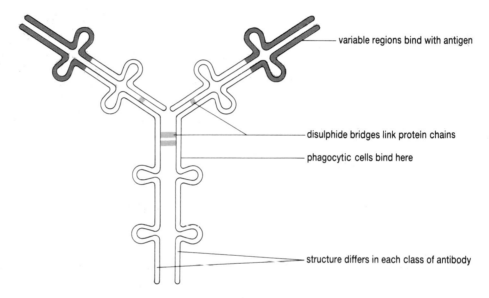

variable regions bind with antigen

disulphide bridges link protein chains

phagocytic cells bind here

structure differs in each class of antibody

FIG 7.3 The specificity of the antibody molecule lies in the two ends of the arms of the Y shape. These sections are unique to each type of antibody and exactly match the surface antigens of the microorganisms they were raised against.

How the immune system can make so many different antibodies

Lymph nodes are small organs located in strategic places around the body. Within them there are clusters of new unspecialised B-lymphocytes, which carry antibody fragments on their surfaces that have the potential to match antigens on foreign material. During the course of an infection, the invading microorganism is presented to these lymphocytes in the lymph nodes. Any B-lymphocytes with antibody fragments matching an antigen on the microorganism are selected and multiply quickly, producing clones of **plasma cells** that can produce large amounts of antibody matching the antigens. Antibody production is slow at first but high concentrations can be detected within a few days of meeting a new antigen. Some of the plasma cells become memory cells, which survive for years ready to initiate another bout of antibody production, should the pathogen return. Each different strain of pathogen carries slightly different versions of its antigens, so new slightly different antibodies from new B-lymphocytes are needed as the pre-existing antibodies will not match.

Different types of antibody molecules are involved in different aspects of defence. One sort is active on mucosal surfaces where it reacts with microorganisms in the airways and gut. Another type passes across the placenta to protect new born babies. Two more help activate complement.

7.4 WHITE CELLS

Several sorts of white cells are involved in the response to microbial infections. Phagocytic white cells engulf particles carrying foreign antigens, and lymphocytes secrete antibody and cell-activating substances. These cells play major roles and differ in appearance, as can be seen in Figure 7.4. White cells called eosinophils tackle parasitic worms and flukes that are too large to be engulfed. They release an enzyme to degrade the parasite's body wall. They are more effective if the worm or fluke is tagged by antibody.

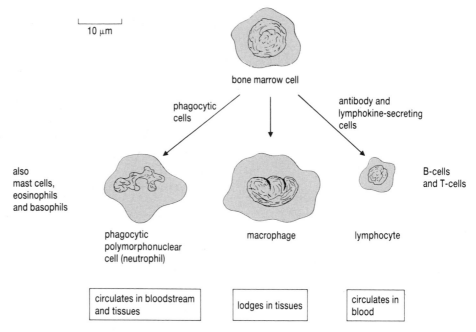

FIG 7.4 White cells.

Phagocytic cells

Phagocytic cells, shown in Figure 7.5, are scavenging white cells in the blood and tissues that pick up living and dead organisms, cell debris and other particulate matter. They respond to a variety of signals found on microorganisms and work better if the material they are scavenging is linked to antibody, complement, or carries certain common bacterial components. There are two different sorts of phagocytic cells. **Polymorphonuclear monocytes**, also known as neutrophils, are the most common white cells circulating in the blood. They can migrate through capillary walls into tissues. They are attracted to the site of damage or infection by chemicals, particularly histamine and small peptides, released by bacterial breakdown, and search for particulate matter to engulf. Receptors on their membranes bind with antibody or complement attached to the material they are scavenging. This stimulates the phagocytic cell to engulf the particle by enclosing it in a vacuole, a process known as endocytosis. Within the cell, lysosomes containing a mixture of hydrolysing enzymes and peroxide-producing enzymes fuse with the vacuole and the microorganism is killed and degraded.

Most bacteria die within half an hour but *Brucella* and *Mycobacterium tuberculosis* bacteria endure these conditions and grow within the cells, causing chronic infections that are difficult to clear up. Bacteria with capsules (see Section 2.2) are more resistant to being engulfed because antibody cannot attach to tag the cell and phagocytic cells are inhibited. Phagocytes are very short-lived but are important in the first response to wounds and infections.

Macrophages are phagocytic cells that are permanently lodged in tissues. They originate as white cells in the blood known as monocytes but they quickly leave the circulatory system and settle in the tissues in areas where organisms could gain entry. They are common in the gut, in lung tissue, in the liver and in the spleen. Macrophages are better able to ingest bacteria if they are activated by T-lymphocytes (see below). Macrophages last longer than phagocytes and are important in deep-seated or chronic infections.

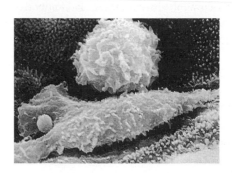

FIG 7.5 The nearer of these two macrophages is in the process of engulfing a particle. The 'arms' will eventually surround the particle enclosing it in a vacuole.

Lymphocytes

Lymphocytes circulate in the blood, and are also found in the lymphatic system, lymph nodes, and interspersed with macrophages in the spleen and in other places. Lymphocytes start out as **stem cells** in the bone marrow, but finish their development elsewhere to become either T- or B-lymphocytes. **B-lymphocytes** are matured in lymph nodes around the body and become antibody-secreting cells in the process outlined above. T-helper cells (see below) activate the process of multiplication to make a clone of antibody-secreting cells. B-lymphocytes migrate from the blood to infection sites, drawn by chemicals released by other cells. They secrete large amounts of antibody as they circulate.

Most lymphocytes are **T-lymphocytes,** called T-cells. New discoveries about T-cell activity and the chemicals they release are still being made, and the overall picture is not yet clear. Collectively they organise and control the immune response using cytokines, which are chemicals they secrete that act as signals between immune system cells, and also activate other cells. T-cells also monitor healthy tissues for cells with signs of abnormality. Any cell that is different, be it a microorganism, a virus-infected cell or a rogue body cell, is targeted for destruction. T-cells are subdivided into a number of groups, each with specific jobs.

T-cell activity begins when a phagocytic monocyte presents antigen fragments from foreign material such as a virus to them. These phagocytic cells in the blood engulf anything carrying foreign antigens. Antigens from the degraded material pass to the phagocytic cell's surface next to a group of important surface proteins called the MHC (major histocompatibility complex). The antigen–MHC complex is presented to T-cells called T4-cells. A T4-cell with a matching receptor for the antigen can then launch a response to that antigen. Once a T-cell recognises, for example, virus proteins among the main histocompatibility antigens, it initiates a cascade of activity, including inflammation reactions. It releases chemicals called interleukins which enhance the immune response and an enzyme that allows T-cells to pass through blood vessel walls to get to damaged tissues.

The roles of the various types of T-cells are still not fully understood but include:

- T4-cells which regulate the activity of the immune system. They switch on B-lymphocytes to make antibody. They recognise foreign antigens. They produce interleukin-2, which enhances T-cell growth.

- T-helper cells secrete lymphokines that activate other immune cells such as B-cells and T-cytotoxic cells. T-helper cells are dedicated to one particular antigen.

- T-suppressor cells damp down the immune response by reducing the antibody response and suppressing T-cytotoxic cells.

- T-cytotoxic cells (killer T-cells) use an enzyme to kill cells with strange surface antigens such as virus infected cells, cells showing virus antigens, and antibody-tagged bacteria. Killer T-cells also destroy faulty cells such as incipient tumour or cancer cells or other cells with strange antigens such as transplanted tissues, unless they are suppressed.

There is much active research into the cytokines released by lymphocytes, with a view to exploiting them as medicines. For example, tumour necrosis factor (TNF) is released by macrophages and other cells. It was discovered to be cytotoxic to cancerous cells but had little effect on normal cells. Its effect is to cause cells in tumours to commit suicide – a process called apoptosis.

QUESTIONS

7.13 Explain why antibodies to a cold you caught early in the year will not prevent another cold virus infecting in the autumn.

7.14 Why do microorganisms die when a phagocytic cell engulfs them?

7.15 Review the section above and draw up a table of white cells and their functions.

7.16 Give one reason why a disease like TB is so difficult for the victim to deal with. After you have worked through the relevant section in Chapter 8, explain why AIDS is also such a problem to the immune system.

7.17 A particular species of pathogenic bacteria was investigated and it was observed that the normal virulent strain had a capsule but that a mutant strain unable to make a capsule could not infect healthy animals. Suggest two explanations for this observation.

7.18 Find out from your core biology textbook more about (i) lysosomes, (ii) the lymphatic system.

7.5 VACCINES AND ANTISERA

As early as the 10th century Chinese and Turkish people were given infections to protect against further infections. In the 18th century Dr Edward Jenner exploited this observation when he used a cowpox infection to protect against smallpox. The term vaccination was coined to describe the process, which became widespread.

An active infection raises an immune response that protects against further infection, but the same immune response can also be raised artificially by **immunisation.** The early vaccinations of the 18th century used live cowpox virus to stimulate an immune response that would respond to similar antigens on smallpox viruses. Nowadays vaccines, which are microbial suspensions, are used to raise an immune response and reduce the incidence of disease. Resistance to an infection acquired in this way is called **active immunity.** Not all diseases can be prevented by vaccines, however; some organisms simply don't provoke a strong immune response. Nevertheless, antibodies may be still be used to treat or prevent an infection, giving **passive immunity.**

Passive immunity and antisera

Passive immunity is provided by giving someone a solution of specific antibodies, or antiserum, to protect against an infection for a short time. Antiserum gives immediate protection whereas it takes several days for a vaccine to provoke the production of large amounts of antibody. The antibodies in **antiserum** reduce the effects of an infection. It lasts for about 3 weeks before it disappears from the system, so it is a short-term measure. Antisera are also useful in treating infections for which there are no really effective drugs. Babies gain passive immunity when they absorb antibodies from their mother's milk, or when particular antibodies cross the placenta into the developing baby.

Antiserum is usually made by injecting an antigen solution into a suitable animal to provoke specific antibody production. Plasma containing the antibodies is collected from the animal and concentrated. Some people react to animal antiserum so it cannot be used for everyone. Human antibodies can be extracted from the blood of people who have recovered from a disease; for many years this was the only way in which Lassa fever could be treated. A general pooled human serum made from the combined plasma of many people can also be made. Now, specific monoclonal antibodies, see Chapter 6, can be made and used to treat particular diseases.

What is a vaccine?

Over 60 vaccines are commonly used in developed countries to give protection, mainly against bacterial and viral diseases. Vaccines contain antigens from a particular microorganism to provoke a lasting immune response in the recipient. Some vaccines use whole organisms. Whole organisms carry the complete range of antigens, and vaccines using whole organisms usually provoke a good response, but several inoculations may be necessary. Vaccines such as whooping cough vaccine are made from killed organisms, whereas the vaccines for polio and rubella use live weakened strains. Weakened, or **attenuated**, strains stimulate the immune system but do not cause a severe infection.

Vaccines such as the tetanus or diphtheria vaccine contain a **toxoid**. A toxoid is an inactivated bacterial toxin that is harmless but still stimulates the immune system. Newer vaccines may contain just purified fragments of organisms, which carry antigens, like the 'flu vaccine. Vaccines are usually injected into the bloodstream, or into the skin layers or muscle. There are a few oral vaccines, such as the polio vaccine, that can withstand passage through the stomach and are absorbed through the gut wall (Figure 7.6).

The newer vaccines, for example hepatitis B vaccine, use genetically modified organisms to make vaccine materials. They carry DNA that encodes proteins made by a pathogenic organism. There are several other possibilities under investigation. One proposed technique is to use a vaccine composed of a live disabled virus such as *Vaccinia* (cowpox), modified to carry the genes for a particular pathogen's antigen, together with a DNA plasmid carrying the same genes. The plasmid ferries the antigen-making genes into host cells. When the genes start to work in the new host cells, their products will stimulate the body's immune system, supported by active genes in the modified virus. The individual receiving the vaccine never comes into contact with the pathogenic microbe, merely some of its products and a disabled virus. You can read more about what is in a vaccine and how it is produced in Section 6.7.

The effectiveness of a vaccine, or its **efficacy**, depends on a number of factors. Some antigens generate a strong immune response that protects the recipient for life; others, like the tetanus vaccine, need periodic boosting. A few pathogens have weak antigens that do not provoke a strong response, for example the malaria

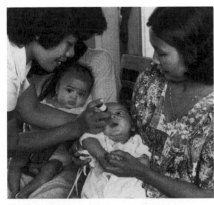

FIG 7.6 Not all vaccines have to be injected. Some, like this polio DPT vaccine, can be absorbed from the gut, which makes vaccination a pleasant experience.

parasite, or the memory cells formed have a shorter life, or the vaccine does not provide protection in a high enough proportion of the population receiving it. Weak antigens may be mixed with **adjuvants** to provoke a stronger response. Pathogens such as 'flu exist in a number of different strains and a vaccine has to carry antigens from all the common strains to be useful.

What happens when you have a vaccination

You can see an outline of the process in Figure 7.7. A **primary response** develops after the first inoculation. Antibody starts to appear in the blood and increases slowly over the next few days. A second inoculation between 4 and 6 weeks later provokes a secondary response. The antibody concentration is quite low after the first injection but rises to very high levels within 48 hours of the second injection, and lasts a long time. A booster dose from time to time maintains the secondary response.

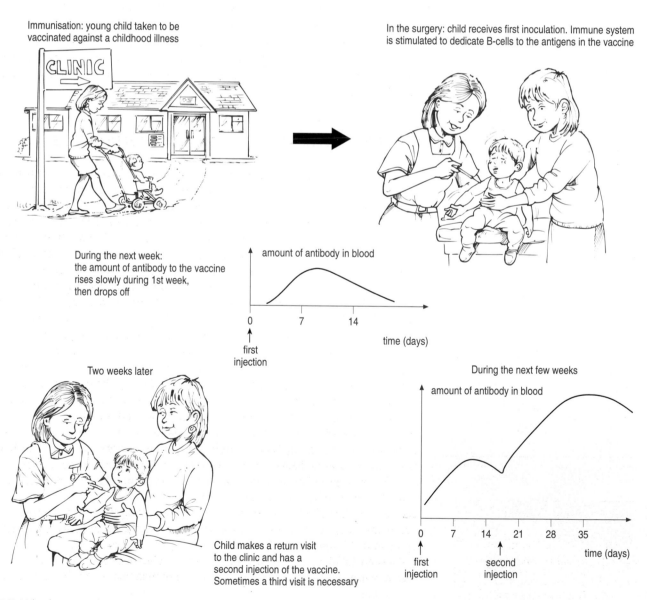

Immunisation: young child taken to be vaccinated against a childhood illness

In the surgery: child receives first inoculation. Immune system is stimulated to dedicate B-cells to the antigens in the vaccine

During the next week: the amount of antibody to the vaccine rises slowly during 1st week, then drops off

amount of antibody in blood

time (days)

first injection

Two weeks later

Child makes a return visit to the clinic and has a second injection of the vaccine. Sometimes a third visit is necessary

During the next few weeks

amount of antibody in blood

time (days)

first injection

second injection

FIG 7.7 What happens when you are vaccinated.

Vaccination programmes

Most countries have a policy of vaccinating young children against the childhood diseases that are still major killers. During the first few months of life, babies are protected against common organisms such as the tetanus bacterium by maternal antibodies, but by 2 months old a baby's immune system can respond effectively to vaccines. The early years immunisation programme includes the following vaccines: DPT against diphtheria, whooping cough (pertussis) and tetanus; polio; MMR (measles, mumps and rubella); and Hib and meningitis C against two forms of meningitis. Although rubella, or German measles, is not particularly troublesome to children or adults if they catch the virus, it is very dangerous to developing embryos. The virus destroys nervous tissue and embryos exposed to the virus during the first 3 months of pregnancy are particularly likely to suffer major damage. Children receive boosters before starting school and at various times during the school years when they are also immunised against TB. This strategy reduces the incidence of these infections, and the numbers suffering severe complications as a result of infection. Nevertheless, millions of children world wide still die each year from measles, tetanus and whooping cough. Even if only a proportion of children are vaccinated, diseases can be controlled if the number of potential victims is low. As long as at least 60% of the population is vaccinated against a disease, it will remain as a sporadic infection, though the target is a much higher proportion. More serious epidemics can occur if the level of community protection falls below 60%.

Why people still catch these diseases

Some children are not vaccinated, even in countries with well-developed primary health services. Some are not vaccinated because they are at particular risk of suffering side effects. A couple of early vaccines contained cell components that caused side effects in a very small proportion of the population, but newer vaccines are made in different ways. The newest vaccines made using modified bacteria contain only the specific antigens. Nevertheless, some parents are concerned about the very small possibility that their child could suffer severe side effects from a vaccine and so do not have their children immunised. There are quantified risks administering vaccines. The vaccines are carefully prepared because of the very small risk that live virulent virus could survive the processing, or that a weakened strain may regain virulence. About one in 300 000 recipients receiving a vaccine using an attenuated strain may suffer a severe infection. These risks have to be balanced against the risk of catching the infection and its major consequences. The vaccine risk is very low compared to the risks in suffering the major disease.

Other children start but do not finish their course of vaccinations for a number of reasons, though vaccinations in school catch many of these. Some children may not be vaccinated because their parents are unaware that their children could suffer; in developed countries people seldom see severe cases of childhood infectious diseases and assume that the infections have died out, or are not likely to infect their own children. Many are unaware of the complications that can occur with what seem to be quite minor diseases.

Over 2 million children world wide die and many more are handicapped each year by measles, a preventable disease. There are still regular measles epidemics in Britain and children do die. The number of cases of measles, and the mortality, in England and Wales over the last few decades is shown in Table 7.2. The use of a measles vaccine was introduced as part of a combined vaccine during that period.

TABLE 7.2 Notifications of and mortality from measles in England and Wales.

YEAR	CASES	DEATHS
1967	460 407	100
1968	236 154	51
1970	307 408	42
1972	145 916	28
1974	109 636	20
1976	55 502	14
1978	124 067	20
1980	139 485	26
1982	94 195	13
1984	62 079	10
1986	82 054	10
1988	86 001	16
1990	13 302	1
1992	10 268	2
1994	16 374	1
1996	5 613	–
1998	3 728	3
1999	2 438	–

Statutory Notification of Infectious Diseases, Public Health Laboratory Services, 2001

More information

Update these figures by visiting the Public Health Laboratory Service:

■ www.phls.co.uk

Can vaccination eliminate a disease?

If there are few enough potential victims in a community for a pathogen to pass to, it will die out in that community. It is possible to eradicate a few diseases completely by vaccination, but only if certain conditions are fulfilled:

- the disease must only infect humans, with no animal reservoir,
- there must be a cheap effective vaccine available,
- immunity must last for life,
- there should be no long-term carriers.

Smallpox was the first disease to be eradicated. As the number of countries with cases of smallpox declined it became feasible to track down all outbreaks and prevent the disease being passed on. Read the Spotlight box on smallpox eradication to find out how it was done. There are only a few diseases that match the criteria above and polio was targeted by the WHO for eradication by the year 2000. However, despite the clearance of the disease from the Americas there are still sporadic outbreaks of polio in many countries. The number of cases reported by 1998 was 80% down from the numbers when the eradication campaign was launched in 1988. Figure 7.8 shows the difference after 10 years of campaign. Measles is another candidate for eradication. Progress on vaccination campaigns can be seen on the WHO website.

More information

- World Health Organisation: www.who.int

Global Polio Situation 1988

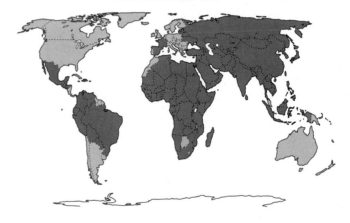

Global Polio Situation 1998

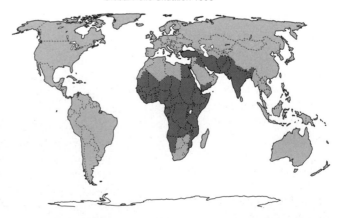

FIG 7.8 How well the polio campaign has succeeded in freeing the world of polio.

■ known or probable wild polio virus transmission
■ zero reported virus but insufficient surveillance and high risk
■ zero reported virus with excellent surveillance or low risk

The eradication of smallpox

So far, smallpox is the only disease that has been completely eliminated from the world's population. Figure 7.9 shows Ali Maow Maalin from Somalia – a remarkable person, he was the world's last case of smallpox. He developed the disease in October 1977 but recovered and worked in primary health care to reduce the incidence of other diseases such as measles, which killed his younger sister. Smallpox was common for hundreds of years; many thousands died and others were permanently scarred. The virus, which does not infect animals, is extremely infectious and causes an extensive skin infection that can even spread to the lining of the gut. The disease needs intensive care and patients have to be isolated when nursed to prevent the disease spreading. Smallpox killed millions of Incas and native Americans when it was taken to the New World with early settlers.

There is a very effective vaccine against smallpox that protected large populations, and the incidence declined as vaccination became common. The World Health Authority decided to focus efforts on a 10 year campaign to eradicate the disease by vaccination, and they began tracking infected people. At the time, they estimated that there were between 10 and 15 million cases of smallpox each year. Health workers visited communities, no matter how remote, and administered vaccine. Any infections they found were treated and the spread of infection was blocked. The last few cases were in Africa in 1977, where just over 3000 people were diagnosed as having smallpox, five in Kenya and the rest in Somalia. Despite a combination of heavy rain and civil war, health workers were able to track down the last few cases. It had taken 10 years, 9 months and 26 days and about $200 million, which was far less than the amount spent annually on controlling and coping with the disease. Every case of chickenpox or other skin rash was examined to ensure that there were no other smallpox sufferers with a minor infection. The eradication was certified in May 1980 and the routine vaccination against the disease ceased.

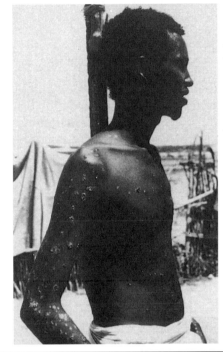

FIG 7.9 Ali Maow Maalin came out in the world's last smallpox spots on 26th October 1977.

QUESTIONS

7.19 Distinguish between active or natural immunity and passive immunisation.

7.20 What is a vaccine? Review Section 6.7 and summarise vaccine manufacture.

7.21 Examine Figure 7.10, which shows the events that occur in the human body over a period of time following the administration of a vaccine such as the typhoid vaccine. Explain the body's response to the administration of the vaccine
(i) in the first two weeks,
(ii) after the second injection.

FIG 7.10 Change in antibody concentration in the blood.

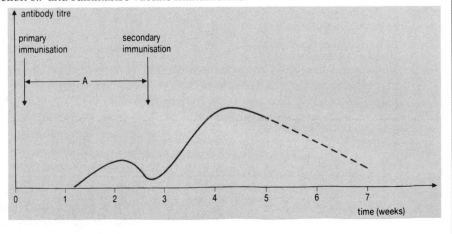

7.22 Examine Table 7.2 and then use the figures to construct a graph of the incidence of measles over the period from 1967 to 1999.

(i) Comment on the figures.

(ii) Is there a correlation between the number of cases and the number of deaths?

7.23 In 1974 there was a health scare about whooping cough vaccine having severe side effects in some children. Figure 7.11 illustrates the incidence of whooping cough between 1940 and 1998. Explain the shape of the graph in sections A, B and C.

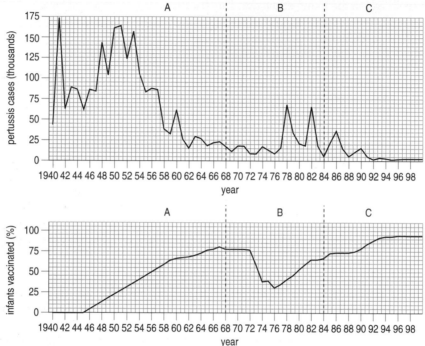

FIG 7.11 The number of cases of whooping cough between 1940 and 1998.

7.24 Find out about Edward Jenner's work developing a vaccine against smallpox. What ethical issues does his work raise? How are vaccines tested today?

7.25 By the end of the 19th century the population of native North Americans had fallen to about one-tenth the numbers when Columbus landed, the majority killed by epidemics of measles, 'flu, chickenpox and whooping cough brought by European settlers. Use your knowledge of the passage of disease and the immune system to explain why native Americans suffered far more than the settlers did.

7.26 List the vaccines you received and the age at which you were inoculated. Compare your list with others in your class. Have you all been inoculated against the same things? Ask older members of your family what they were vaccinated against. Why is there a difference? What other vaccines would you expect to receive if you were a young child today?

7.27 Measles is caused by a virus which spreads from person to person like 'flu. Explain why the incidence of measles in small communities of fewer than 5000 people is low. Measles is much more common in large populations. Explain why a vaccination policy gives a high level of protection even when some people are not vaccinated.

7.6 DRUGS AND ANTIBIOTICS

Chemotherapy is the use of drugs to treat disease. Traditional remedies used plant extracts and other ingredients to treat illnesses. Many natural medicines and traditional remedies have been investigated and the active ingredient identified. Other drugs have developed as a result of a scientific understanding of how a microorganism causes disease. Research into the structure of drugs and the biochemistry of drug action has changed the way we use them. For instance, the bark of the cinchona tree was used over 300 years ago to treat malaria. Then its active ingredient was extracted for use; now we use cheaper synthetic derivatives of quinine with enhanced activity. Table 7.3 lists some chemotherapeutic agents and their uses.

TABLE 7.3 Chemotherapeutic agents

NAME/CHEMICAL TYPE	ACTIVITY/USE
erythromycin	antibiotic, inhibits 50S ribosome subunit
penicillin, a ß-lactam	antibiotic, blocks cell wall synthesis in some bacteria
nystatin	antibiotic with anti-fungal effect in animals, from *Streptomyces noursei*
sulphonamides, e.g. sulphamethoxazole	analogue to 4-aminobenzoic acid, a coenzyme needed in bacterial metabolism
tetracycline	antibiotic from *Streptomyces aureofaciens*, prevents tRNA linking to ribosomes
amphotericin B	anti-yeast activity in animals
pentamidine	anti-protozoal activity, treats sleeping sickness
streptomycin	antibiotic from *Streptomyces griseus* with anti-TB activity, causes misreading of mRNA, incorrect amino acid incorporation in protein molecules
chloramphenicol	antibiotic which blocks peptide bond synthesis
rifampicin	antibiotic, inhibits mRNA synthesis

Drugs

At the turn of the 20th century there were no effective drugs to cure common infections, and though disinfection had been discovered it was not widely used, so many patients died of diseases and septic infections. The discovery of synthetic drugs came with the work of Paul Erlich who was scanning newly synthesised chemicals for any that could kill pathogenic bacteria. In 1909 Erlich and Hata worked on an organo-arsenic compound they had synthesised which was effective against syphilis bacteria. It was the first synthetic drug for people, though it did have side effects. Once one chemical had been found that killed bacteria but did not kill the victim, others were sought. Within a few years sulphonamide drugs were developed and many more have been discovered since.

The action of penicillin, an antibiotic, was first observed shortly afterwards, though it was not in use medically until 15 years after the first observations. A few years after penicillin came into general use, streptomycin then tetracycline were discovered, then a number of others. Anti-viral agents were not discovered till much later and there are still relatively few effective anti-viral drugs because anything that prevents virus multiplication will actually affect some part of normal human cell activity and will damage the host too. Scientists looked for compounds with anti-microbial activity almost randomly at first but now the search for new drugs is more systematic. Knowing what happens to cells during an infection enables us to find chemicals that will interfere with the activity of perhaps just one enzyme. Knowing the chemical structure of the active site of a drug enables researchers to construct computer models of molecules that will be more effective, or have fewer side effects, or improved physical properties. These can then be synthesised and tested. Even so, developing a drug is expensive and time-consuming, costing millions to develop and taking 10 years before it can be marketed. Few promising compounds ever reach the pharmacy because many are too expensive to make, or are too toxic to use.

Antibiotics

An antibiotic is a substance produced by a living organism that kills or inhibits the growth of other microorganisms. Antibiotic activity was first observed scientifically in 1928, when Alexander Fleming observed that a fungus, *Penicillium notatum*, inhibited the growth of bacteria on an agar plate. Many years later the compound responsible was isolated and identified, then eventually small amounts of penicillin were obtained. Human trials took place in the early 1940s and it was brought into general use after 1945. Thousands of antibiotics have been discovered, but relatively few are used as medicines. Many are too toxic to use, or they are not as effective as compounds already in use, or they are uneconomic to bring into production. You can read about penicillin manufacture in Section 6.8.

Most antibiotics work by inhibiting a biochemical pathway in the target microorganism, blocking its growth. Some aspects of metabolism are unique to microorganisms, so an antibiotic targeting these should not adversely affect human or animal cells. Antibiotics affecting processes common to animals and microorganisms, such as tetracyclines which affect protein synthesis, have to be used carefully. Some people do suffer adverse reactions to antibiotics. For instance, a significant proportion of the population is allergic to penicillin and its derivatives. Most antibiotics work on bacteria but there are a few anti-fungal compounds.

An antibiotic will affect only those bacteria that have the particular property targeted by it. For instance, penicillin, a β-lactam, inhibits the synthesis of peptidoglycans in new bacterial cell walls, resulting in a loss of strength and eventually osmotic lysis. Only the Gram-positive bacteria, with significant amounts of peptidoglycan, are affected by penicillin. Antibiotics effective against a wide range of organisms, for example tetracyclines and chloramphenicol, are described as **broad spectrum**. Those affecting just a few species are **narrow spectrum**.

Drug and antibiotic resistance

Within a population of bacteria, individuals will differ slightly from each other, just as the human population varies. Occasionally selection pressures in the environment make some of the differences important. Someone with a bacterial infection is host to hundreds of thousands of bacteria with random mutations in many of their genes. One of these mutations might confer resistance to a drug's mode of action. Microorganisms will be resistant to a drug if they have an alternative pathway available to them, or evolve one. Penicillin resistance is due to an enzyme, β-lactamase, that degrades penicillin. The genes for β-lactamase are carried on a plasmid within resistant bacteria. The mutation has no special advantage to the organism when the drug is not present; in fact it might even be disadvantageous.

Unfortunately treating infections with drugs selects **for** those organisms with resistance mechanisms. These survive the treatment and continue to multiply, but their fellow invaders are killed or inhibited. Body defences mop up some resistant microorganisms but others continue to cause infections. They may spread to other victims and will very quickly come to dominate the local population. The more frequently and widely a drug is used, the stronger the selection pressure for resistance. For instance, widespread use of aminoglycoside antibiotics to treat *Staphylococcus aureus* and *Pseudomonas aeruginosa* infections, which cause sepsis, has led to resistance. In Austria by 1985 42% of strains of *S. aureus* were resistant to the common aminoglycosides such as gentamicin. In contrast, when it was realised what was happening, the use of these drugs was restricted in Scandinavia where resistance settled at 2–3%.

Resistant strains of bacteria such as staphylococci, pneumonococci, salmonella and mycobacter are now common. These are dealt with by using two or three drugs combined together, each acting in a different way. Although a strain of pathogen may be resistant to one drug, it is unlikely to be resistant to three different drugs acting in three different ways simultaneously. Leprosy is treated with dapsone together with two other drugs, rifampicin and clofazimine, as a triple drug following the appearance of strains resistant to dapsone in many parts of the world. However, the rise of multi-drug resistant strains has become a major issue in hospital care. Bacterial plasmids carrying resistance genes are freely transferred between related bacterial species. This phenomenon is known as **transferable drug resistance**.

The practice of giving farm animals antibiotics as a routine preventative has also encouraged drug resistance. Bacteria with resistance genes on a plasmid in (say) a chicken's gut can spread along food chains and supply lines and transfer its plasmid freely to many other organisms.

QUESTIONS

7.28 **(a)** What is the difference between a broad spectrum and a narrow spectrum antibiotic?

(b) Give two ways in which antibiotics can reduce bacterial growth.

7.29 Explain how a population of bacteria can develop resistance to an antibiotic. Name one thing that can reduce the chance of resistance developing. What is transferable drug resistance and why is it a problem?

7.30 Alexander Fleming first observed the effect of an antibiotic when he noticed that bacteria did not grow around the site of penicillin production. How could you use this effect to compare the effect of the antibiotic streptomycin on the growth of two different bacteria?

Exam question

7.31 *Haemophilus influenza* type b (Hib) is the bacterium that causes Hib meningitis. A carbohydrate from the bacterial coat was used to prepare a vaccine against the disease. However, this vaccine was not effective in children under 2 years of age because they produced very few antibodies and were unable to produce the necessary memory cells. When the carbohydrate was combined with a protein from the bacterial coat, a vaccine effective in children less than a year old was produced. This vaccine was introduced into the UK in October 1992. In 1993 the number of cases of Hib meningitis occurring in children under 1 year old was only 25% of that predicted. All these occurred in unvaccinated children. The goal is now the elimination of Hib meningitis from the UK by vaccinating at least 90% of children under 2 years of age.

(a) Explain the function of the protein from the bacterial coat when the vaccine is injected into a child. (2)

(b) Explain how Hib meningitis may be eliminated even though every child is not vaccinated. (2)

(c) Using the information in the passage, explain why the vaccine containing only carbohydrate did **not** provide effective protection against Hib meningitis. (1)

(d) Give **two** other methods used to prepare vaccines. (2)

AQA Biology, March 1999, Paper BY08, Q. 2

SUMMARY

A pathogen is a microorganism that causes disease. All groups of microorganisms have pathogenic members. Pathogens may cause cell and tissue damage, release toxins or affect host cell activity, reducing the ability of the host to survive. Bacteria and viruses are most important in animal disease.

Humans have natural barriers that hinder the entry of pathogens, including the skin and anti-microbial substances in body fluids.

Within the body the immune system produces a response specific to the particular pathogen and provides lasting immunity. The immune response involves phagocytic cells, T- and B-lymphocytes and antibodies. Phagocytes ingest and degrade particulate matter. Antibodies bind to microbial cells and disable them; they act as markers too. Antibodies are specific molecules. They combine with only one type of antigen molecule.

Some pathogens can disable host defences by a variety of mechanisms.

Active immunity is raised during an infection or as a result of vaccination. It stimulates the body's own defence mechanisms. Passive immunity is the result of the donation of antibodies from another source.

A large proportion of the population needs to be vaccinated in order to break the transmission chain.

Drugs are available to treat microbial infections. Antibiotics are effective against bacteria. Resistant bacteria are becoming common.

HUMAN DISEASE

LEARNING OBJECTIVES

After studying this chapter you should be able to:

① explain how disease-causing microorganisms are transmitted

② appreciate how the incidence of disease can be reduced

③ understand the causes of a range of diseases

④ give detailed accounts of selected diseases and their control.

8.1 INTRODUCTION

The focus of this chapter is human diseases caused by microorganisms. We are infected by pathogens from most microbial groups, particularly bacteria and viruses. Very few of the thousands of known bacterial species are harmful, but those that are have played an important part in human affairs. We have only to think of plague bringing economic change to Europe in the Middle Ages, or the constant drain caused by diseases such as cholera and TB in many parts of the world. Viruses almost always disrupt normal cell activity and cause illness. Protozoa cause few diseases (malaria is one) but they are globally important. There are few serious fungal diseases but even trivial infections can be very damaging to people who are debilitated.

Each kind of microorganism is considered separately, with a brief survey of the problems they cause, and a few serious diseases are covered in more detail. You will need to look at your specifications carefully. The parasitic worms and flukes are not included; though they cause globally important diseases, they are beyond the scope of this book.

Infectious, or **communicable**, diseases pass from one person to another. The microorganisms causing them are the **inductive agents**. Organisms differ in their severity and they are not equally infectious. An organism's ability to cause disease is called **virulence**. Smallpox is very virulent because just a few organisms are needed to start the disease and they cause severe effects within a short space of time. Leprosy bacteria and salmonella food poisoning bacteria have low virulence; many encounters or huge numbers of organisms are needed to cause disease.

Epidemiology

Epidemiology is the study of the incidence and spread of disease, and the factors that influence its spread. The numbers and sources of important infections are monitored and recorded centrally. This will pick up any unusual clusters of infection, or infections in particular age groups or employment categories, or any unusual rise in numbers. The information is used to identify sources, investigate the infection process, identify risk factors, to try to limit the spread of infections, and to prepare health resources for predicted surges in the numbers of infections.

Certain diseases are **notifiable**, that is, any cases are reported to a central monitoring unit. Here disease patterns are carefully monitored, and investigations undertaken to identify sources of infection. Sudden significant outbreaks of diseases are tackled by measures to break the transmission chain. For instance, a local outbreak of meningitis may necessitate antibiotic treatment for the victims' families and other contacts, and a local vaccination campaign. A large food poisoning outbreak leads to a careful investigation into the food and drink consumed by the victims, and microbiological testing of the foods to identify sources of the bacteria. The victims are isolated to stop the infection spreading. Once the source has been tracked down, hygiene measures can be taken to prevent further spread.

Infectious microorganisms circulate in all populations. However, a particular pathogen is only found where there are enough potential victims to keep its population going, and where transmission from one victim to another is relatively easy. Measles virus was unknown on many Pacific islands because the populations were not large enough to keep the virus circulating. When European travellers brought it to the islands, the virus infected a large proportion of the population, none of whom had any immunity. An **endemic** disease is one normally found in a geographical area. Populations do not necessarily suffer a great deal from endemic diseases; for instance, plague is endemic in wild animals in North America but is seldom transmitted to people.

A few people in a small area contracting an infection form a sporadic outbreak. An **epidemic** involves large numbers of people suffering from the same strain of microorganism in many communities. But when very large numbers of people spread across large geographical areas and several countries suffer from the same infection, it is a **pandemic**. There were pandemics of measles and other diseases across North America when European settlers arrived, and worldwide pandemics of 'flu several times during the 20th century. The most recent human pandemics have been of 'flu and AIDS. It has been said that 'flu killed more people in the 1918 pandemic than were killed in the First World War, and HIV is infecting over a quarter of the population in some countries – a scale similar to the Black Death.

How do we know what is causing the disease?

An important factor in dealing with any infectious disease is identifying the causative agent. Simply finding an organism growing in an infected individual is not enough to say that it is the causative agent. Damage done to tissues during an infection provides an opening for any microorganism that happens to be there at the time and can exploit the situation. Robert Koch did much of the pioneering work identifying the organisms that cause particular diseases and he laid down a set of criteria, **Koch's postulates**, which are still useful when trying to establish causative agents:

- the microorganism must be observed in all cases of the disease,
- it must be isolated in pure culture from diseased tissue,
- microorganisms from the pure culture must reproduce the disease in animals,
- the organism must be recovered from the experimental infection.

Despite the hundred years since Koch began identifying causative agents, there are still infections whose causative agent is uncertain. Environmental factors and genetic influences that affect the progress of infections within different individuals complicate matters. For example, why do so many people exposed to TB bacteria not go on to develop infections? Why do most people clear chickenpox from their systems but others develop shingles later in life? Where did HIV come from? Epidemiology answers some of the questions as patterns are revealed when large numbers of cases are studied, but much remains to be investigated.

QUESTIONS

8.1 Write definitions of the following: communicable disease, endemic disease, inductive agent.

8.2 What is the difference between an epidemic and a pandemic?

8.3 What is meant by a virulent strain of bacteria?

8.4 Find out what you can about Dr Snow's work establishing the cause of a cholera outbreak in London in the 19th century, and breaking the transmission chain. Use your material to prepare a short talk for your fellow students.

8.5 Visit the Public Health Laboratory Service website to investigate disease incidence and vaccination programmes.

More information

- Public Health Laboratory Service: www.phls.co.uk

HOW MICROORGANISMS ARE TRANSMITTED

8.3

Pathogenic microorganisms can spread to new hosts from one of several reservoirs of infection.

- Many diseases are infectious before there are any obvious symptoms, so individuals with active infections can pass microorganisms to their contacts or to their local environment.

- Individuals who have had infections such as typhoid and hepatitis but who are not completely clear of the organisms may become **symptomless carriers**. They excrete the pathogen but do not show any signs of infection. Carriers need to be identified and treated to clear the infection if possible.

- Some animals harbour pathogenic microorganisms; a disease which infects both people and animals is called a **zoonosis**. We share many diseases with other mammals, for example bovine TB infects cows, badgers and humans. Control measures have to deal with the pathogen in animals and in humans. This is particularly difficult if the reservoir is a wild animal.

■ Many pathogens are found in soil or water. Water-borne pathogens are easier to control with water purification but soil-borne pathogens such as tetanus bacteria are much more difficult.

Microorganisms rely mainly on passive transmission from one person to another, but some exploit **vectors** such as insects to carry them from one host to the next. A few species are mobile, generally using flagella to swim through moisture films, but the distances they may have to travel from one host to another are huge compared to their size. They still largely depend on chance contacts. The time spent in transit from one host to another is dangerous for the microorganisms because the temperatures may be lower than their optimum if they are adapted to life inside mammals, and they are subject to drying and UV light. The length of time that an organism can survive outside its host varies; those passed by direct contact have a very short survival time but others may be transmitted as resistant structures such as spores. Transmission of microorganisms can be divided into a number of broad categories, though some organisms can be put into more than one category.

Direct contact. Diseases can be transmitted by direct contact between an infected individual and someone else. The mucous membranes lining the nose and mouth and genital tracts are thinner and softer than most outer surfaces so they are the most frequent entry points, though several organisms infect through the skin. Direct contact is the main means of transfer for most sexually transmitted diseases such as gonorrhoea.

Directly into the blood. This is an uncommon method of transfer but the few diseases spread this way are severe. Transfer occurs when blood or body fluids from one individual enter the bloodstream of another. Usually this happens when a secretion such as seminal fluid or saliva comes into contact with a damaged mucous membrane, allowing entry directly into capillaries. Pregnant women may pass infections across the placenta to their developing child. Sexual activity is a common means of transfer and so is the practice of sharing hypodermic needles among drug abusers.

Wounds and bites. Commensals living on the skin surface enter through broken skin and start an opportunistic infection. For instance, staphylococci living on the skin may infect tissues to cause boils or occasionally septicaemia. Tetanus bacteria live in the soil, degrading dead organic material, but dirt can carry spores into deep wounds, or they can enter via an animal bite.

Droplet transmission. Organisms that infect the respiratory tract are transmitted in air currents. They are carried in exhaled air or in the fine spray of droplets of saliva and mucus expelled by coughs, sneezes or running noses. Large droplets settle quickly; they dry up and the pathogen is released as a fine dust. Smaller droplets stay airborne until the moisture evaporates, leaving small particles called droplet nuclei. These take a long time to settle and are easily inhaled by another person.

Water-borne and food-borne. Many pathogens are transmitted via food or water. The pathogens can get into food at source, for example gut bacteria in carcass meat, or are brought by flies and other organisms touching food. *Salmonella* bacteria are passed on by handling contaminated raw meat or poultry or eating undercooked food. You can find out more about *Salmonella* in food in Chapter 4 and the disease it causes in Section 8.4. Some pathogens enter during food preparation as a result of human failure to observe basic rules of hygiene, such as washing hands after visiting the toilet. These organisms infect through the gut and often leave via the gut, or cause diarrhoea. Typhoid and cholera bacteria are transmitted in faeces. Inadequate sewage treatment leads to water becoming contaminated with organisms.

Vectors. Vectors transmit many viruses and harmful protozoa. The most important vectors are insects, including biting flies and bugs that transfer pathogens as they feed on blood. Biting flies, like the one in Figure 8.1, have mouth parts adapted to penetrate through skin. Salivary secretions that prevent blood clotting as the insect feeds carry the pathogen into the victim. Protozoal pathogens may live for part of the life cycle within the vector, so it has to be adapted for two different environments. For example, the plague bacterium affects rats and humans but is carried by fleas. Similarly, the malaria parasite is adapted for life in the human body as well as an *Anopheles* mosquito's digestive system. You can read more about malaria in Section 8.6.

FIG 8.1 The tsetse fly, *Glossina*, feeds on blood and may carry parasitic protozoa in its salivary glands. The parasites are *Trypanosomas* species, which cause sleeping sickness. The parasite's presence in the feeding parts can interfere with the insect's ability to monitor blood flow while feeding, causing it to feed more frequently, and for longer, assisting the parasite's chances of transfer.

TABLE 8.1 Selected diseases and their means of transmission

DISEASE	INDUCTIVE AGENT	TRANSMISSION
athlete's foot	*Trichophyton rubrum* (F)	soil-borne and water-borne spores
leptospirosis (Weil's disease)	*Leptospira interogans* (B)	wound entry from contaminated water, infects man and rodents
hepatitis B	hepatitis virus (V)	via blood
whooping cough	*Bordatella pertussis* (B)	airborne droplets
measles	measles virus (V)	inhalation of virus
polio	poliomyelitis virus (V)	drinking water
rabies	rabies virus (V)	vector into blood
sleeping sickness	trypanosome (P)	vector, bite by tsetse fly
syphilis	*Treponema pallidum* (B)	direct contact in sexual activity
typhoid	*Salmonella typhi* (B)	faecal contamination of food and water

Note: B = bacterium, F = fungus, P = protozoa, V= virus.

QUESTIONS

8.6 Write definitions of 'zoonosis' and 'carrier'?

8.7 List three measures that can be taken by a community to promote the health of its members.

8.8 Give one example of a pathogen from each of the main groups of disease causing microorganisms and its means of transmission.

8.9 Why do some organisms need a vector to transfer them?

8.10 Look at the information in Table 8.2. It gives the average incidence of subjective travellers' diarrhoea in British package holiday tourists visiting popular holiday destinations during the 1996 summer season.

(a) What do you think 'subjective travellers' diarrhoea' means?

(b) What other things might you wish to know about the questionnaire before you could reach firm decisions about the data that it contains?

Assuming the data to be fair:

(c) What trends can you see in the data?

(d) Most travellers' diarrhoea is caused by faecal microorganisms in water that is used for drinking, ice, washing food and brushing teeth. Design a simple bookmark-sized good hygiene reminder, with six points to remember, that could be slipped into the traveller's ticket package by the travel agent.

TABLE 8.2

HOLIDAY DESTINATION	% SUBJECTIVE TRAVELLERS' DIARRHOEA	NUMBER OF QUESTIONNAIRES ANALYSED
Egypt	63	13 134
Dominican Republic	57	31 218
Kenya	56	3 469
Mexico	49	3 502
Turkey, Ionian coast	43	2 750
Antigua	22	1 612
Costa Dorada, Spain	16	26 306
South Tenerife	13	21 660
Rhodes	12	25 619
Greek mainland	12	978

Extracted from *World Health Statistics Quarterly*, 50 (1997).

8.4 BACTERIAL DISEASES

Bacteria cause many diseases but most are curable, given appropriate medical resources and access to trained staff early in infection. Table 8.3 summarises the bacterial pathogens that you have met in the previous chapters.

TABLE 8.3 Summary of bacterial diseases mentioned in the book and their causative organisms

DISEASE	ORGANISM
diphtheria	*Corynebacterium diphtheriae*
TB	*Mycobacterium tuberculosis*
leprosy	*Mycobacterium leprae*
anthrax	*Bacillus anthracis*
tetanus	*Clostridium tetani*
whooping cough	*Bordatella pertussis*
brucellosis	*Brucella* spp
bubonic plague	*Yersinia pestis*
cholera	*Vibrio cholerae*
syphilis	*Treponema pallidum*
Rocky Mountain spotted fever	*Rickettsia rickettsii*

In a developed society the bacterial infections that we are likely to meet are not necessarily the ones that cause concern on a global scale. Globally, diseases such as tetanus, whooping cough and diarrhoea take a huge toll on life where vaccinations and good hygiene are less easily available. In Britain people are most likely to meet streptococci, staphylococci, and the enteric bacteria. Most of these are common commensals and are not controlled by vaccination.

Diarrhoeal diseases

Between 5 and 10% of the UK population has the misfortune to suffer an infection causing diarrhoea each year. In healthy people these infections result in an uncomfortable and embarrassing day but for 3 million under-5s each year it is the cause of their death.

The infection generally known as food poisoning is caused by a number of different enteric bacteria as well as diarrhoea-causing viruses, such as rotaviruses and hepatitis A. Cholera, dysentery and typhoid are more serious infections affecting the gastrointestinal tract. These serious infections often accompany wars and civil strife when thousands of people flee their homes and gather in refugee camps. Where large numbers of people gather in a small area with limited supplies of clean water and poor sanitation, cholera outbreaks quickly begin.

Salmonella food poisoning

The organism

Most cases of food poisoning are caused by *Salmonella* bacteria, including *S. typhimurium* and *S. enteriditis* phage type 4, but others such as *S. virchow* are often recorded. *Shigella* and increasingly *Campylobacter* and *E. coli* O157 also account for a number of cases. Together *Salmonella* account for half the food poisoning cases reported in Britain. They are flagellate rod shaped bacteria, shown in Figure 8.2, that are part of the normal gut flora of many farm animals, especially intensively reared animals. Rats and mice also carry the bacteria. An increasing number of *Salmonella* are finding their way into unpasteurised milk and eggs, hence into foods made with raw eggs. You can read more about *Salmonella*, *E. coli* O157 and food in Chapter 4.

The bacteria are not particularly virulent; we have to take in about 10 million live salmonella before an infection starts. This is quite different to *Salmonella typhi* where just a few organisms are needed to start a typhoid infection. Bacteria are confined to the gut of farm animals; meat and poultry become contaminated after slaughter. Meat is chilled or frozen straight after slaughter so there is little bacterial growth until defrosting. Bacteria can pass from defrosting meat to other foods through films of moisture, or via hands or utensils used in preparation. Most reported cases arise from meat that has not been properly cooked and is left in warm surroundings, or from food that has come into contact with bacteria.

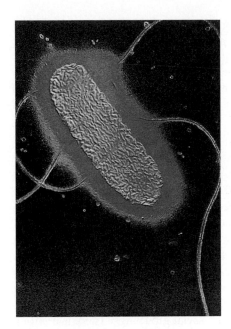

FIG 8.2 *Salmonella*, responsible for a large proportion of cases of food poisoning, can swim through moisture films to contaminate food or water.

The disease

Symptoms begin a few hours after eating infected food, though it may take a couple of days. The patient has abdominal pains, feels feverish, suffers diarrhoea and may be sick. This can go on for several days, resulting in an enormous loss of fluid and electrolytes from the body. Bacteria colonise the cells lining the small intestine, where the endotoxin has damaging effects. They multiply within these cells, kill them and provoke an inflammatory response. Damage to the microvilli on the intestinal mucosa leads to absorption and possibly osmotic problems, and problems with digestive enzyme production. Very young or malnourished children, elderly people and people already suffering other infections can suffer a fatal fluid loss.

Treatment and control

Antibiotics are not very effective because it is difficult to get enough into the cells where the bacteria cause damage. The main treatment is care of the patient, and in severe cases a sterile solution of glucose and electrolytes is fed as a drip into a vein to replace lost fluid. Most patients recover within a week.

The disease is controlled by improving the standards of hygiene at farms, at meat processing plants, and at all stages of food handling and preparation. It is difficult to stop chickens acquiring the bacteria from faeces in the deep litter in chicken houses, or from feed contaminated by wild bird droppings, and chickens pass it to each other. Some chickens develop a carrier state and can infect others. Flocks are regularly monitored for the presence of salmonella. In some countries young chicks are vaccinated against *S. enteriditis* before they begin to lay. Food handlers have to be aware of the hazard posed by uncooked meat and have to follow regulations concerned with minimising the growth or transmission of organisms.

Cholera

Cholera is a very serious epidemic infection caused by *Vibrio cholerae* in food or water. It frequently accompanies disasters where people have to live in primitive conditions with little sanitation. The bacteria multiply in the upper part of the small intestine for several days before the typical violent purging diarrhoea starts. The patient produces a huge volume of very watery diarrhoea. Over half of untreated patients die.

The bacteria do not invade the body but remain within the lumen of the intestine where they produce a toxin – verotoxin. The toxin damages mucosal cells lining the small intestine. Water and salts are lost from the blood and tissue fluid, resulting in dehydration and an ion imbalance, particularly acidosis from the loss of hydrogen carbonate ions and consequently nerve and heart problems from the loss of potassium ions. The loss of so much fluid and salts from the body is life threatening. Treatment involves replacing the lost fluids with sterile rehydrating solution through a sterile intravenous drip.

Vaccines are not very effective. The organisms do not enter the bloodstream or the tissue lining the gut to any great extent so they do not come into contact with antibodies. The most effective strategy is to have a water supply and disposal system that keeps human sewage away from sources of drinking water, and hygiene by food handlers.

Tuberculosis

Tuberculosis was very common a hundred years ago, but a vaccination programme in the 1950s and the discovery of an effective drug treatment brought it under control in Britain and many other parts of the world. However, TB has spread again, even in affluent countries, with about 20 million cases world wide. About 3 million people die each year, estimated to be about a quarter of all adult deaths in developing countries. A huge number of people have dormant infections – they are infected but are not ill. The disease becomes active when they catch something else, particularly HIV or other infections affecting the immune system.

The incidence differs markedly between social groupings. There is a strong correlation between decreasing standard of living and increasing deaths from TB. There is a higher incidence in countries where there are poor people with little access to health care. Migrant workers and economic immigrants from poorer countries suffer more TB than indigenous populations. Though bovine TB has also increased, it is not the same strain of organism that causes the human rise.

The organism

TB is caused by a bacterium, *Mycobacterium tuberculosis*. It is one of the organisms whose whole genome is known, but the function of many of the genes remains unknown. It grows slowly in laboratories making it awkward to research. The bacteria die quickly when exposed to ultraviolet light, but can stay alive for long periods in dark, dank conditions.

The disease

About 10% of people infected by the bacterium develop an active disease. We do not fully understand why the infection is trivial and unnoticed in so many people but is fatal for others.

The human strain of TB usually affects the lungs and spreads in mucus droplets when sufferers cough or sneeze. People can live in the same household as a sufferer without contracting the infection for long periods. The infection begins as a small inflammation near the top of the lungs where TB bacteria infect macrophages. Bacteria are usually destroyed by macrophages (you can read more about this process in Chapter 7), but the TB bacteria can resist intracellular digestion in some people's macrophages. Even if the bacteria survive, active infection does not always follow. A proportion of infected people limit bacterial spread within a granuloma, which is a mass of tissue that stops the bacteria spreading.

If the infection becomes active, the sufferer loses weight and wastes away – hence the old name of consumption.

The bovine strain is usually spread in infected cow's milk, and the bacteria spread from the intestine to many other body tissues.

Treatment and control

The tuberculin test detects exposure to TB bacteria. An extract of proteins from the TB bacterium is inoculated into the skin. Anyone who has, or has had, an infection will have sensitised T-cells and develops a large red swelling at the test site when they respond to the bacterial antigens. People who have had the BCG vaccine respond similarly and a laboratory test is needed to confirm the presence of TB bacteria.

Infections are usually treated with the antibiotic rifampicin and a thiosemicarbazone drug, isoniazid. People with active infections are given a combination of antibiotics for about 6 months, or longer courses of other drugs, which cures most cases. Once the course of antibiotics is started, people cease to be infectious, but these drugs must be taken daily for months. As many sufferers are amongst the poorest, the most disadvantaged, and with worst access to health care, they may not complete the course of drugs. Even people with good access to health care may stop taking the drugs when they start to feel better. However, the infection has not been beaten at that stage. They continue to carry infectious bacteria and may even have colonies of drug-resistant strains from the limited exposure to appropriate drugs. Resistant strains of TB appeared during the 1980s, including some multi-resistant strains that are virtually untreatable.

Vaccines

The BCG vaccination, which uses a vaccine derived from a weakened related mycobacterium, provokes a lasting immune response. This vaccine controlled the incidence of TB effectively for many years. However, it does not seem to be as useful when used for adults in some tropical areas, so new versions are under investigation.

QUESTIONS

8.11 Before there were suitable drugs, TB patients were treated in 'sanitoria' – away from housing areas with plenty of bed rest in the fresh air and sunshine. Give two reasons why this may have been effective in reducing the incidence of TB.

8.12 Choose one bacterial disease and draw up a table to link the time sequence of the infection suffered by the patient with the activity of the bacterium.

8.13 The chief medical officer for the town of Greater Hartwell has been notified of 15 patients taken ill with food poisoning. Nine had attended a dinner at a local hotel, one worked washing up in the hotel kitchens, two others were members of the family of a guest at the dinner. The menu that evening was prawn cocktail with home-made mayonnaise, steak with jacket potatoes and salad, and raspberry gateau to follow.
(a) List the likely sources of the infection.
(b) What steps should the chief officer take to
(i) identify the source or sources of infection,
(ii) prevent further spread of the disease?

8.5 VIRAL DISEASES

Viral infections are more difficult to treat than bacterial diseases because there are fewer anti-viral drugs available. Any drug that interferes with virus nucleic acid or protein synthesis will usually affect the patient's own cell processes that the virus has hijacked. Vaccines have reduced the incidence of many viral diseases such as smallpox and polio. Table 8.4 lists some important viral pathogens. The most common viral infections are caused by the Rhinoviruses that cause the common cold. There are many different varieties unfortunately and immunity is short-lived, so there is little hope in the near future of preventing colds.

TABLE 8.4 Summary of viral diseases mentioned in the book

DISEASE	ORGANISM
infectious mononucleosis (glandular fever)	Epstein-Barr virus
chickenpox	*Herpes varicella*
smallpox	variola
polio	poliomyelitis virus
pneumonia	cytomegalo virus
AIDS	human immunodeficiency virus
cold sores	*Herpes simplex type I*
yellow fever	togavirus

The *Herpes* viruses are also common. They cause diseases such as cold sores, chickenpox and viral pneumonia. Herpes viruses are unusual because in an infection some migrate through neurones to the central nervous system. There the virus integrates into the DNA of neurones, making a **latent** infection. Years later, stress or another stimulus causes the virus to emerge and travel down the neurone to cause problems again. For example, people with cold sores caused by *Herpes simplex type I* may only need to go out walking on a bright cold winter's day to reactivate the infection. Similarly, some children with chickenpox, caused by *Herpes varicella-zoster*, can suffer later from shingles, which are painful blisters in the skin served by the nerve with the latent virus.

Influenza

Every winter influenza, or 'flu, circulates in the population, but every few years we suffer an epidemic, and at longer intervals a pandemic. Healthy people recover from a 'flu infection after 10 days or so but thousands of very young, old or run-down people die during epidemics, particularly if complications such as pneumonia set in. The 'flu figures for England and Wales can be seen in Figure 8.3. The disease also causes a huge economic loss through people losing time from work. Several strains infect people but only type A causes severe pandemics; Type B is less serious. Type A can infect a wide range of mammals and birds too.

FIG 8.3 The incidence of 'flu. Large peaks in the graph indicate epidemics.

The disease

'Flu is caused by a myxovirus. It has RNA as its genetic material, but it is unusual because it is in eight strands instead of one. 'Flu viruses can be seen in Figure 8.4. The virus particle is surrounded by an envelope of host membrane but two types of protein molecules in the virus capsid project through. Haemagglutinin is used to attach to receptors on respiratory tract cells, and induces the cell to take in the virus. After the virus has replicated itself, new viruses bud through the host cell membrane. Neuraminidase is an enzyme that changes cell membranes, so viruses can leave one cell to pass to another. It removes traces of cell receptors from the virus envelope and stops viruses clumping together. The virus is transmitted by droplet infection and infects the cells lining the nose, throat and airways. By 3 days after infection the viruses are causing serious damage to the mucous membranes. Bacteria such as staphylococci may cause secondary infections in damaged tissues. Interferons released by lymphocytes protect some cells from infection but also cause aches, pains, chills and a runny nose. Viruses are transmitted to other people in droplets of moisture from the lungs.

Treatment and control

'Flu is generally treated by rest, and drugs to relieve the worst symptoms. There are no drug cures for 'flu but patients who are very seriously ill may receive some anti-viral drugs. Good care reduces the chances of secondary infections that would need antibiotic treatment. Keeping the patient at home reduces the chances of transmitting the virus to other people.

Antibodies are formed to the antigenic virus coat proteins. These proteins change subtly year by year and each winter new variants of the main 'flu strains circulate. Pre-existing antibodies may not be effective against the new variant and vaccines prepared against earlier strains may not work well. 'Flu virus also undergoes major changes periodically, with variations from animal versions of 'flu. The epidemic of Hong Kong 'flu in 1968 was caused by a virus which had two avian 'flu genes; the 1957 strain had three avian strain genes. It is suspected that a domestic animal, a pig perhaps, comes into contact with human and avian 'flu strains and acts as a mixing pot, allowing new variants to develop. The fragmented nature of its RNA allows greater mixing. A lethal variant containing avian genes developed in Hong Kong in 1997 but it was contained when all poultry infected with avian 'flu were slaughtered.

New strains can be recombined with a fast growing cultured strain to produce new vaccine in about 6 months. People most at risk, that is, medical staff, the elderly and those with respiratory problems, are protected by yearly vaccinations with the latest vaccine. This prevents about 70% of infections and reduces the numbers of elderly patients hospitalised or dying.

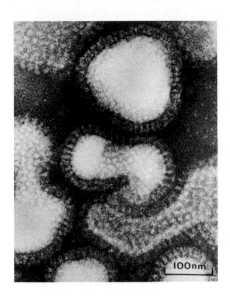

FIG 8.4 Influenza particles may have an irregular shape as well as having the genome in several sections. The spikes enable it to infect cells.

AIDS

HIV, a retrovirus, causes AIDS, or acquired immune deficiency syndrome. HIV contains two strands of RNA as its genetic material, which is an uncommon form of genome. Its structure can be seen in Figure 2.14. AIDS first started to become common in the 1980s, although earlier sporadic cases of AIDS-like conditions have been identified in retrospect. It is widespread through central Africa, parts of North America, Europe and the Far East. By 1999 it was estimated that 34 million people were infected; thousands are infected weekly and millions have died. 70% of cases are in sub-Saharan Africa and are a significant proportion of the population.

HIV is passed from person to person by blood and body fluids. There are three main ways in which it is transmitted: in intimate contacts such as sexual intercourse; in blood through blood transfusions or shared hypodermic needles; and it can pass to babies across the placenta or during childbirth.

What happens during an infection?

Infection by the HIV virus leads to the destruction of a large part of the immune system. Soon after infection, large numbers of virus particles can be found in lymph nodes. The virus binds to specific CD4 receptors on T-helper cells, a kind of lymphocyte, and some other cells in lymph nodes (more about these cells in Chapter 7). T-helper cells help regulate the immune system's activity by recognising foreign antigens in the system and by switching on B-lymphocytes to make antibodies. The virus core enters the cell where an enzyme it carries, reverse transcriptase, synthesises DNA from its RNA. This DNA is inserted into the lymphocyte's DNA and copies remain in the DNA for life.

The virus often lies **latent** for a long time before it begins to replicate. Virus proteins seem to prevent infected cells carrying out their usual functions. Only a few T-cells appear infected at any particular time, but large numbers of them gradually lose their ability to respond to foreign antigens and die. New viruses within a T-cell bud off from the cell membrane, weakening the cell which eventually ruptures within a couple of days. Other mechanisms seem to lead to the death of helper T-cells too. Numbers in the blood fall from about 800 per cm^3 to less than 200. Large numbers of immune system cells linked to the lymph nodes are destroyed. Circulating virus-laden monocytes take the virus to many parts of the body, safe from the immune system's activity.

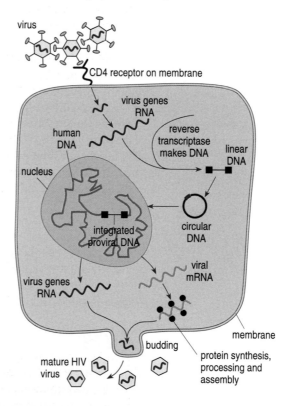

FIG 8.5 The HIV infection cycle. Several years may elapse after infection before the virus begins to replicate in large numbers.

The average incubation period is about 10 years. Patients suffer fairly generalised conditions such as fever, diarrhoea and weight loss. They also suffer from opportunistic infections such as thrush or toxoplasmosis that are much more severe than normal. Later there may be more severe conditions such as Karposi's Sarcoma – an uncommon cancer, and dementia from infection directly into brain cells. Effectively, HIV destroys the body's ability to fight off infections and deal with rogue cells, and many AIDS victims die of opportunist infections. In developing countries many AIDS victims die of TB. HIV virus exists in strains and it seems to be able to change its antigens subtly so the depleted immune system cannot cope with it or generate enough new cells to replace those killed.

Treatment and control

There is no cure for AIDS but there are drugs that slow down the development of the infection. Drugs such as AZT (zidovirudine) inhibit the action of reverse transcriptase, whereas protease inhibitors block a viral protease enzyme that HIV needs to multiply. As neither of these enzymes is normally found in human cells they should not affect healthy cells. The drugs have side effects that can be severe, and patients are treated with combinations of the two types of drugs to suit their state of health and viral resistance. The virus has evolved resistance to some of the drugs, so a combination therapy is less likely to develop resistant strains. The drugs can significantly reduce the risk of the virus passing through the placenta to an unborn child. If the mother also delivers by caesarean section and bottle-feeds, the child is likely to be free of infection.

There is no successful vaccine at the moment. Trials with vaccines made from dead viruses or extracts containing viral proteins have not provoked an immune response. The virus is very variable and there are several strains of the virus. It has recently been discovered that people can become infected with two different strains at the same time.

The spread of HIV depends on intimate contact between individuals and so breaking the transmission chain relies on people making an effort to avoid becoming infected or passing on the infection. People with the infection need to know how to avoid passing it to others. World wide, the main transmission route is through sexual intercourse, so people are encouraged to use barrier contraceptives such as condoms, which are very effective at preventing the virus passing from one person to another. However, in many badly hit regions people do not have easy access to information, health care or condoms, and whole families suffer as children contract the infection from their mothers. One of the biggest challenges is getting people to change their behaviour and cultural norms to reduce transmission.

QUESTIONS

8.14 Why are virus diseases mainly controlled by vaccinations?

8.15 Which people are most likely to benefit from 'flu vaccination?

8.16 Examine Figure 8.3 which shows deaths from 'flu. Describe the general trend of 'flu deaths during the 1900s. Give three reasons for the change in numbers. In which years do you think there were epidemics?

8.17 Reread Section 7.5 and explain how antiserum is useful for treating a virus infection.

8.18 What do you think is meant by a virus 'lying latent'? Give one example of a latent virus.

8.19 Reverse transcriptase from retroviruses has an important role in biotechnology. Reread Section 6.5 to find out what it is used for.

There are few pathogenic protozoa, but they cause very serious problems, particularly in tropical areas and developing countries. Many pathogenic protozoa can live in other animals and may be transmitted by insect vectors, so vector control is as important as control of the protozoan. A few, such as dysentery, are transmitted by contaminated food or water. Table 8.5 lists some of the important protozoal diseases. The *Plasmodium* species, causing malaria, and the *Leishmania* group, which causes a whole range of diseases, are responsible for illness in hundreds of millions of people world wide.

TABLE 8.5 Protozoal diseases

DISEASE	PATHOGEN	NATURE	TRANSMISSION
amoebic dysentery	*Entamoeba histolytica*	invades gut mucosa	water-borne
Kala Azar	*Leishmania donovani*	generalised infection; joints, liver, spleen, immune system all affected	carried by sandflies
trichomoniasis	*Trichomonas vaginalis*	vaginal and reproductive tract infection	sexual transfer
Chagas disease	*Trypanosomas cruzi*	fever, infection of heart and other organs	carried by bugs
sleeping sickness	*Trypanosomas brucei*	blood infection	carried by Isetse fly

Malaria

Over two hundred million people are infected with malaria at any given time, and about two million die each year. Pregnant women, with depressed immune activity, and just-weaned children lacking protection by maternal antibody in milk, are particularly at risk. Malaria is a collection of infections caused by four closely related protozoa of the genus *Plasmodium*. These are *Plasmodium vivax*, *P. falciparum*, *P. malariae* and *P. ovale*. The dominant form in Africa is *P. falciparum*, which is spreading rapidly and causes the most severe illness. Each protozoan species has a specific *Anopheles* mosquito vector that transmits it when it takes a blood meal. Though malaria is most common in the African countries south of the Sahara, nearly 150 countries world wide support the *Anopheles* mosquitoes. Malaria has been eradicated from some of these countries and an enormous effort is being made to control or eradicate the disease in the remaining countries.

Plasmodia are very successful parasites as they can survive and evade human immune responses, multiply rapidly, and evolve resistance to most chemotherapeutic agents. They are well adapted to life in animals as different as mosquitoes and man. They also modify the behaviour of the vector in ways that increase the likelihood of transmission. For example, infected flies spend more time probing before feeding and probe more often. This arouses the victim, who is more likely to disturb the fly. The fly has to attempt to feed on more hosts to get the blood it needs, thus increasing the chances of transmission.

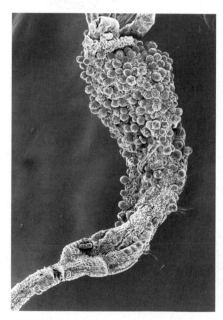

FIG 8.6 The malaria protozoan occurs in several different forms during its life cycle. The surface chemicals vary too and are not strongly antigenic.

The disease

The *Plasmodium* life cycle is shown in Figure 8.7. Sufferers have bouts of severe fever every few days, together with anaemia and liver and spleen damage. The parasite lodges inside red blood cells where it uses haemoglobin as a source of amino acids. It changes cell membrane activity to take in huge amounts of glucose from plasma and it generates large amounts of lactic acid and haem. The protozoa reproduce rapidly within red blood cells and a bout of fever accompanies their release with all their wastes to infect further red cells. The frequency of bouts depends on the species that is causing the infection. In some species, parasites in the liver can remain dormant instead of dividing, and form a focus of reinfection months later.

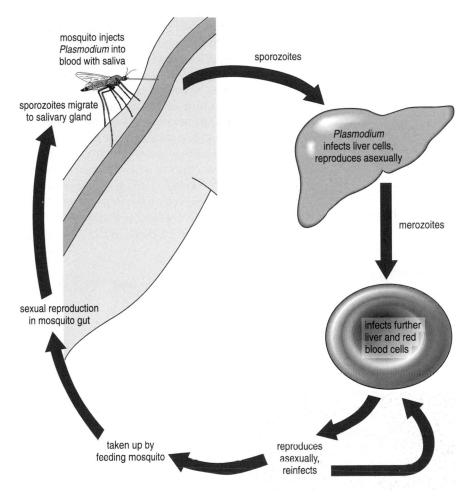

FIG 8.7 The life cycle of the malaria parasite.

A lot of tissue damage is caused by the body's response to the infection. Fever temperatures and immune system responses to tissues that harbour protozoal antigens damage the kidneys and other organs. Aggregations of damaged cells and parasites block blood vessels, which then causes damage to tissues beyond the blockage. This is particularly dangerous in the blood vessels supplying the brain. Many patients recover after a few bouts of fever, but anaemia and exhaustion take their toll. If the victim is malnourished or has other infections, malaria can accelerate death from other causes.

Control and treatment

Malaria is managed by a three-part strategy:

■ drugs to prevent infections becoming established,

■ treating infected people,

■ controlling the vector.

Once the mosquito has injected the protozoa they quickly disappear from the bloodstream into cells. There is little opportunity for the immune system or drugs to stop the infection establishing itself. Once in liver and red blood cells, the parasites are safe from the immune system, but malaria can be cured if it is diagnosed and treated promptly. Some drugs act against blood stages, others against liver stages, but none is effective against every stage. Drugs such as chloroquine kill blood stage parasites. The most important preventative measure is to keep a high enough concentration of a suitable drug in the blood at all times. This kills the parasites when mosquitoes first inject them. Promaquine and pyrimethamine are also used.

All the major drugs are based on quinine, and resistance to molecules based on its shape has arisen. Chloroquine resistance has spread around the world and it fails to control malaria in many regions. In some places there are multi-drug resistant strains. Figure 8.8 shows the spread of drug resistance. Drug-resistant strains are treated with mefloquine or halofatrine. Mefloquine, the most active drug, is combined with two other drugs, sulphadoxine and pyrimethamine. The three drugs are **synergistic**, that is they enhance each other's activity, so they are more effective when combined than on their own. Though mefloquine resistance can develop, resistance to a combined treatment is less likely; an organism would have to evolve three different resistance mechanisms at the same time. Some people react adversely to mefloquine but if travellers start taking the drugs 3 weeks before the risk of infection, any adverse reaction will have appeared and the drugs can be changed.

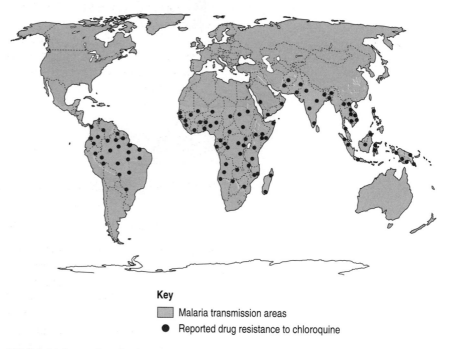

Key

▨ Malaria transmission areas

● Reported drug resistance to chloroquine

FIG 8.8 Malaria is found in hotter countries world wide. The parasite has evolved resistance and this is spreading world wide too.

Artemesin, extracted from *Artemesia* plants, has anti-malarial activity and has an entirely different structure, so it could be useful against resistant strains. Artemesin releases toxic free radicals when it comes into contact with iron in parasitised red blood cells. Another approach under investigation removes surplus iron from the blood. The parasite needs iron for an enzyme involved in DNA production. Without sufficient DNA, reproduction stops.

There is no effective vaccine yet. Antibodies made during infections do not seem to prevent reinfections or relapses. However, a patient who has survived several infections does eventually become immune. Different antigens appear at various stages of the life cycle, and the parasite seemingly can change its surface antigens enough to reduce the effectiveness of pre-existing antibodies. Also, there is an enormous variation in antigens between strains. Vaccines containing a selection of antigens from different stages have been tried, but because the blood stage antigens are specific to each strain. an effective vaccine would have to include antigens to many strains. There is also research on DNA vaccines (see Chapter 6).

Controlling the mosquito vector

Several approaches to mosquito control are used, often in combination.

■ Dark places where mosquitoes rest during the day are sprayed with insecticide. This is effective but mosquitoes have evolved resistance to most agricultural insecticides and there are problems associated with insecticide residues.

■ Mosquitoes breed in still water; they can even breed in water collected in a tyre rut after it has rained. Covering water in cisterns and tanks with a lid stops mosquitoes gaining access for egg laying. The larvae hang from the surface tension and breathe air. Detergent and oil sprayed on open water reduce surface tension and the larvae drown.

■ Mosquitoes are attracted to the smell of sweaty feet. Washing and covering up feet and ankles and arms in the evening when mosquitoes fly discourages bites.

■ People can substantially reduce their chances of being infected by using insecticide-treated mosquito nets over their beds, insect repellents, screening the doors and windows of their houses, and burning mosquito coils. Trials of bed nets treated with pyrethroid insecticides in west Africa – where people are bitten by mosquitoes 10 times more often than elsewhere – have had tremendous success in reducing the death rate. However, treated bed nets and the cost of twice yearly re-treatment is very expensive compared to the local family income.

QUESTIONS

8.20 Briefly explain why a vaccination policy is not very effective in controlling malaria.

8.21 Describe three ways of controlling the spread of malaria.

8.22 There were 2364 reported cases of malaria in the UK in 1997, with 13 deaths. 1401 were falciparum malaria cases, up from 1283 the previous year and 1113 the year before that. Falciparum malaria is mainly contracted in east and west Africa. Write a newspaper article aimed at people planning to visit relatives in east or west Africa about how to reduce the chances of catching malaria during the visit. You should include advice about preventative drugs and how to prevent mosquito bites.

8.23 Using the diseases influenza, AIDS and tuberculosis, draw up a table giving the type of microorganism causing the infection and one method of reducing the incidence.

Exam questions

8.24 **(a)** Describe the circumstances when people are most at risk of becoming infected with cholera. (6)

(b) Discuss the reasons why smallpox was eradicated by vaccination, but so far there is no successful vaccination for cholera. (10)

OCR Sciences, November 1999, Paper 4808, Section B, Q. 2

8.25 The graph shows some of the changes in the body after infection with Human Immunodeficiency Virus (HIV).

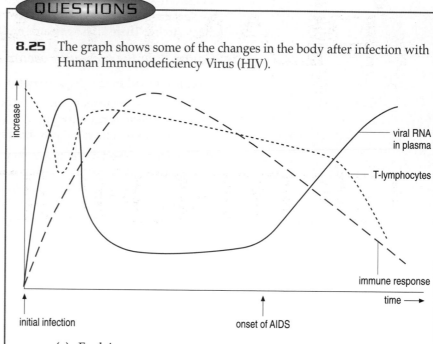

(a) Explain
   (i) the decrease in the number of T-lymphocytes after initial infection, (1)
   (ii) the low concentration of viral RNA in the plasma between initial infection and the onset of AIDS. (1)
(b) Use the information in the graph to explain why patients are more prone to common infections in the later stages of development of the disease. (2)
(c) Mutation of the Human Immunodeficiency Virus reduces the ability of the immune system to destroy the virus during the early stages of infection. Mutations are less likely to occur if drugs which inhibit viral replication are used soon after infection.
   (i) Explain why mutation of HIV reduces the ability of the immune system to destroy the virus. (2)
   (ii) Explain why mutations of the virus are less likely to occur when these drugs are used. (1)

AQA Biology, June 1999, Paper BY08, Q. 6

SUMMARY

A pathogen is a microorganism that causes disease. Its virulence is its ability to cause disease.

Viruses, bacteria and protozoa cause the most important human diseases.

Pathogenic organisms are spread by direct contact, through food and water, by droplets, bites, wounds or blood to blood contact, or by vectors.

Infectious diseases are controlled by treating infected individuals with appropriate drugs, by preventing infection by vaccination or drug treatment, and by environmental management to break the transmission chain. Diseases spread by vectors are controlled by reducing the numbers of vectors. Integrated strategies involving all aspects must be used. Economic factors such as lack of access to health care or cultural factors can hinder efforts to reduce disease incidence.

Some infections have no easy cure, so control rests with prevention.

Globally important diseases include TB, AIDS, malaria and diarrhoeal diseases.

9 PLANT DISEASE AND BIOTECHNOLOGY

LEARNING OBJECTIVES

After studying this chapter you should be able to:

① describe the main causes of plant disease

② understand why infected plants have a reduced yield

③ explain how plant pathogens are transmitted

④ give a detailed description of selected plant diseases

⑤ describe the control methods used against plant disease

⑥ understand why plants have modified genomes

⑦ explain how plants can be propagated using biotechnology.

9.1 PLANT DISEASE

Plants are the foundation of all food chains, and getting the best productivity from crops is vital to the growing human population. Plants provide other useful products including pharmaceuticals, construction materials, fabrics, perfumes, hygiene products, pigments and paper, among others. Microbiologists and biotechnologists work on reducing the effects of pests and diseases on plant growth, improving the yield of plant products and generating large numbers of plants for agriculture and horticulture.

Fungal spores were identified shortly after microscopes came into use and the role of fungi in decay was recognised. Mercury compounds were used to preserve wood, and fungicides came into use in the 19th century. The outbreak of potato blight in Europe in the 19th century focused attention on plant disease, and the link between the blight and a fungus was finally established.

It is difficult to assess how much damage plant infections cause but mildew on cereals alone is thought to cause several hundred million pounds of lost crop value.

9.2 MICROORGANISMS THAT CAUSE PLANT DISEASE

Most plant pathogens depend on having a living host plant, and debilitate rather than kill. Fungi are the most important; over 1000 fungal species have been found associated with plant disease. Viruses are also important pathogens and there are some plant diseases that are caused by agents even smaller than viruses called satellite viruses and viroids (see Table 2.3). Bacteria cause few problems because they cannot push through the cuticle of leaves and stems or plant tissues, they cannot move far around the plant, and a plant's acidic interior does not suit them. The red rust that parasitises citrus plants is caused by algae but algal pathogens are unusual.

9.3 HOW PLANT DISEASES ARE SPREAD

Most crop infections are the result of spread from other infected crops or from infected wild plants growing at the edges of fields. The infection may be quite innocuous in wild plants – they have evolved resistance mechanisms over thousands of years – but have major effects in crops. Crop plants have been selectively bred to improve yield, and genes for disease resistance are often lost in the process.

Crop plants are grown in parts of the world far from their origins and meet new pathogens against which they have no resistance. Pathogens spread too, with seeds, tubers, vegetables, fruit and other plant products exported from one country to another. The plant populations of their new host country may be unable to resist the infection and it spreads quickly.

Farming practices such as **monoculture,** which is the growth of a single strain or **cultivar** of a crop plant over a large area, like that shown in Figure 9.1, make it easier for pathogens to spread. Crop plants are bred to be genetically consistent and a pathogen that can initiate an infection will reproduce very quickly asexually. This generates large numbers of genetically consistent pathogens that spread rapidly through acres of virtually identical crop plants, causing enormous damage. As farmers tend to grow the same few commercial crop varieties, the infection can spread over large areas.

This section concentrates on fungi and viruses, which are the most common plant pathogens. You may find it helpful to read about the biology of fungi and viruses in Chapter 2 first. Some fungal pathogens belong to the normal microbial community found on leaves, bark and around roots. They cause **opportunistic infections** when they unexpectedly gain entry to the interior of a plant through surface wounds caused by wind damage, biting insects or browsing herbivores. Plants suffer particularly from **soil-borne** fungal diseases. The fungus may grow in the soil or lie dormant until a suitable plant grows nearby, then it infects.

Both fungi and viruses rely on external agents to carry them from one plant to another. Fungal spores spread in **air currents** and **water splashes**. They may be

FIG 9.1 One cultivar of grain covers hectares of land. Any plant pathogen that can infect one plant in this acreage could infect the lot and have devastating effects on the harvest.

FIG 9.2 Feeding aphids push stylets between cells and secrete enzymes to aid the passage. The flow of salivary juices carries viruses with it.

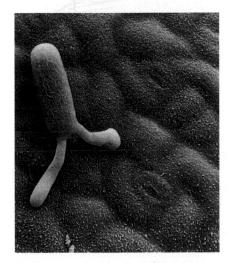

FIG 9.3 The germinating spores of fungal pathogens can either push hyphae through stomata or use enzymes and force to pass between plant cells.

FIG 9.4 Viral activity stops chlorophyll synthesis, resulting in patches of pale cells called a mosaic.

carried hundreds of kilometres in air currents, even across oceans. A few viruses are transmitted by **direct contact**, including tobacco mosaic, potato Y and carnation virus, all of which are particularly contagious. Fungal spores and viruses can be carried to a susceptible plant by **vectors**, such as insects, nematode worms, or even by horticultural tools, wheels, or boots.

Viruses are spread mainly by vectors. Insects are the most important vectors but some viruses are spread by soil-living nematodes that eat plant roots, and by soil fungi. Biting flies, bugs and aphids transfer pathogens when they feed on plant sap. Aphids are the chief culprits; the peach–potato aphid, *Myzus persicae*, is outstanding, recorded as carrying over 70 different plant viruses. Aphids have stylet mouth parts that penetrate the cuticle and epidermis to slip between cells into individual phloem vessels. Figure 9.2 shows an aphid feeding. Aphids feed on the contents of a single cell, so viruses in a cell of an infected plant will stick to the mouth parts. When the aphid feeds on another plant, the viruses are carried into the plant with secretions such as cellulases and pectinases that help the aphid feed. Some viruses have a close relationship with the insect, and even multiply inside their host, which remains infective for the rest of its life. These relationships are specific, each virus multiplying inside a specific host.

When fungi or viruses establish an infection they cause physical damage and drain the plants of materials made from photosynthesis, and may disrupt other physiological processes. Loss of sugars reduces crop yield.

Some pathogens are more damaging than others. There are highly specialised fungal parasites that drain a plant's nutrient stores but cause little other damage, but less well-adapted pathogens severely derange plant activity. Infections can start in above-ground parts, or fungal spores and bacteria in the soil infect through roots. Once the outer layers of the plant are damaged, organisms that could not infect an undamaged plant get in through the wounds and set up infections. Figure 9.3 shows a fungal infection peg pushing through a leaf surface.

Once inside there are several types of damage:

- **Impeded transpiration**: Fungi growing in xylem and phloem interfere with the uptake and transport of water and minerals.

- **Tissue degradation**: Pathogens such as *Erwinia* secrete pectinases and cellulases to break down the plant tissues and release nutrients.

- **Physiological changes**: Many fungi can alter a plant's physiology to provide themselves with nutrients. Infected leaves keep materials made in photosynthesis instead of translocating them to other parts of the plant. Materials are diverted from other parts of the plant into the infected leaf. The two processes make infected leaves appear healthy and vigorous, but the change in carbohydrate storage significantly weakens the plant.

- **Reduced photosynthesis**: Viruses replicating in plant cells divert cell resources away from normal cell activity such as chlorophyll production. This reduces photosynthesis and hence crop yield.

- **Toxins**: A few pathogens make toxins that affect nearby cells or are carried through the transpiration stream.

- **Distorted growth**: Crown gall (shown in Figure 9.8) is caused by *Agrobacterium tumefaciens*, which provokes cell proliferation and makes a large unsightly mass. You can read more about this very important bacterium and its role in genetic modification in Chapter 6. Other infections cause distortion as a result of unusual cell division, such as peach leaf curl, or by interfering with the balance of plant growth regulators.

QUESTIONS

9.1 What is meant by the term 'monoculture'? Why does monoculture encourage the spread of disease?

9.2 Give three different ways in which crop plants may become infected.

9.3 Reread the section above, then construct a table to show how plant pathogens are spread. Use the headings 'wind and water' 'soil-borne', 'mechanical spread' and 'vector'. When you have read the sections on specific diseases, add examples of each method of transfer.

9.4 Give three symptoms you might see that would indicate that your favourite houseplant had a fungal or viral infection.

9.5 How could a pathogen growing in xylem vessels in the stem of a plant cause damage to a leaf?

9.6 Briefly describe two ways that microorganisms can cause damage in plants.

9.7 An investigation of red spot disease, a fungal disease in a particular plant species, was carried out. The plants had all been grown from the same batch of seed in identical conditions. Two of them were inoculated with spores of the disease-causing fungus. Each was illuminated and fed with radioactive labelled carbon dioxide ($^{14}CO_2$) through one leaf, then the distribution of the labelled carbon was monitored 6 hours later. Figure 9.5 shows the distribution of radioactivity.

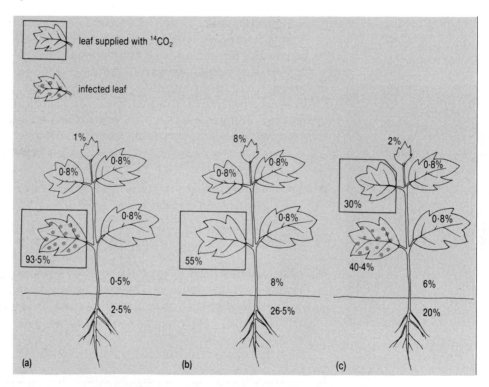

FIG 9.5

(i) How does the fungus affect the distribution of material made in photosynthesis?
(ii) Plant (b) acts as a control for both infected plants. What would have been a better control for plant (c)?

9.4 PLANT DEFENCE MECHANISMS

Plants have evolved defences and natural barriers against pathogenic microorganisms. Some plants contain chemicals that adversely affect invading pathogens; others synthesise substances that inhibit microbial activity once an infection has started. Microorganisms normally found on the surface of a plant, called the phyllosphere flora, are well adapted for life on the surface and germinating fungal spores have to compete with them for scarce nutrients leached from the leaf.

The waxy cuticle is an effective barrier because it may be too thick for fungi to push through, or it may contain toxic chemicals such as gallic acid and terpenes that inhibit the growth of pathogens.

Plants respond to compounds called **elicitors** exuded by invading pathogens. Plants may synthesise substances such as lignin and suberin near the infection to make a physical barrier that blocks the pathogen's spread. Plants may secrete gums to plug xylem and phloem vessels, which stops fungal or virus spread in sap streams. New transport vessels may differentiate to replace the blocked ones. Plants can increase the calcium content of cell walls, making them more resistant to pectolytic enzymes secreted by bacteria.

Resistant plants may have high levels of anti-microbial compounds in their cells. Phenol-based compounds such as lignin precursors, catechol and saponins inactivate fungal enzymes and toxins. Tannins also inhibit virus multiplication.

Phytoalexins are compounds with anti-fungal and anti-bacterial activity formed by plants as a response to cell damage. They are produced in minute quantities by cells near the site of infection but they are not specific to one pathogen as antibodies are.

QUESTIONS

9.8 Explain how the surface of a leaf protects against infection.

9.9 Write definitions of (a) phytoalexin, (b) an elicitor.

9.10 Name two chemicals found in plant cells that may protect against plant pathogens. Give an example of a compound that may be more abundant as a result of a pathogen infecting a plant.

9.5 FUNGAL DISEASES

Fungi are the most important plant pathogens, particularly of crops such as cereals and other staples. In warm and wet weather fungi grow and produce spores in a very short time. Fungal spores are often very resistant to drying and ultraviolet light, resulting in rapid spread of infection. Table 9.1 lists some important pathogens.

TABLE 9.1 Some fungal pathogens

PATHOGEN	PLANTS INFECTED	PROBLEM CAUSED
Armellariella mellea	trees and shrubs	honey fungus, heart and root rot
Botrytis cinerea	many plants	grey mould, destroys tissues, die back
Claviceps purpurea	cereals	ergot, grain infected
Diplocarpon rosae	roses	black spot, leaf damage
Erisyphe graminis	cereals	powdery mildew, reduces photosynthesis and yield
Puccinia striiformis	cereals	rust, reduces photosynthesis
Pythium	seedlings	damping off, seed rots in soil or stem attacked

Many fungal pathogens are opportunistic, living independently in the soil on dead plant material but infecting damaged or ageing plants. *Penicillium* and *Rhizopus*, for example, are both fungi that cannot infect healthy fruit but will grow in wounded apples and cause fruit rot. These **necrotrophic** fungi are unspecialised and infect a variety of plants. Their enzyme action and toxins quickly disrupt the plant's metabolism causing death. In contrast, *Verticillium* and *Fusarium* need living plants for their life cycles. Eventually, however, they kill their hosts and feed on the remains before infecting another plant.

Biotrophic fungi are well adapted to their host plant, slowly draining it of nutrients but not killing it. If no suitable host is growing nearby, they remain as spores in the soil until one is available. Biotrophs usually infect through natural openings such as stomata, and may have highly sensitive growth responses to enable them to locate their entry points. They enter healthy cells to obtain nutrients without eliciting normal responses to infection. Most use a structure called a haustorium, which is inserted through the cell wall into a living cell, pushing between the wall and the protoplast. It alters membrane transport mechanisms to divert nutrients from the plant cell into the haustorium.

Potato blight

Late blight of potatoes is caused by *Phytophthera infestans*. Cultivated potatoes had no resistance to *Phytophthera* at one time and epidemics occurred throughout Britain and Europe. Main crop potatoes are particularly affected; a severe infection can cause the loss of all the leaves on a crop.

The disease

The source of infection is usually infected material left over from a previous crop or planting infected potatoes. The fungus grows in leaves but affects tubers too. The infection begins as a few scattered spots of fungal growth on the undersides of the leaves. The mycelium grows quickly, affecting then killing all the above-ground parts of the plant. The fungus produces huge numbers of asexual spores that are released and spread in rain splashes or by the wind to other plants in the crop. Spores falling to the ground infect tubers near the surface. Infected tubers are slightly discoloured on the outside but have a brown rot inside. Figure 9.6 shows the consequences of the infection. Secondary bacterial infections can cause even greater losses.

Treatment and control

The disease is much worse in wet weather because warm damp weather favours rapid fungal growth. In the UK, agricultural advisers issue a warning about potato blight when the mean temperature has not dropped below 10 °C and the relative humidity has stayed above 75%, sometimes going over 89%, for at least 2 days running. Fungicides such as the dithiocarbamates are used to treat it. There are other fungicides but the pathogen is developing resistance to them. Spraying is repeated regularly because the fungicide is washed off by rain, and new leaf growth is not protected.

Resistant strains and less susceptible strains have been bred. Most strains of early potato varieties are not resistant but they are harvested before the pathogen becomes a serious pest. The transmission chain is broken by good hygiene.

FIG 9.6 Potatoes carrying potato blight or leaf roll virus may appear perfectly healthy on the outside but be brown and discoloured on the inside.

QUESTIONS

9.11   Outline a procedure you could use to investigate the effectiveness of copper sulphate mixed with ammonium carbonate solution (Cheshunt compound) on the growth of a fungus.

9.12   Construct a chart that outlines the potato blight infection cycle from a spore arriving on a plant to the next generation of spores arriving on a plant.

9.6   VIRAL DISEASES

All the major plant groups suffer virus infections, sometimes several different viruses at once. Some infections remain localised in a particular tissue or leaf but in a systemic infection the virus spreads through the whole plant and causes the most severe diseases. Virus diseases are important in agriculture because there are fewer control measures than for fungi.

Virus infections result in paler patches by the veins of an infected leaf, followed by yellow spots as plant cells cease to make chlorophyll. Systemic infections produce more obvious symptoms such as stunting, yellow mosaic patches, distorted leaves, pale streaks in flowers and ring marks on the leaves. Eventually the loss of chlorophyll turns a leaf completely yellow and cells die, causing brown necrotic patches. Viral activity in cells redirects the cell machinery to making more virus, which may reach a large proportion of a cell's mass. The plant's growth slows and the yield of the crop is reduced.

FIG 9.7 Flashes of colour in parrot tulips are induced by viruses; not all virus infections are unwanted!

Potato leaf roll virus

This is a particularly important disease caused by a virus transmitted by *Myzus persicae*, the peach–potato aphid. In the potato life cycle, potatoes produce small tubers in their first year, which overwinter. These seed potatoes are used by farmers to grow for a second year to produce the potato crop. If virus-infected seed potatoes are planted, there is a major yield loss in the second year of growth.

The disease

Aphids transfer the infection from plant to plant. Symptoms are slow to develop; there may not even be any symptoms in the first year. It is not easy to distinguish a virus-infected potato at planting time without cutting it open. The infection becomes apparent when the leaves appear tough and the edges start to roll on the top part of the plant; they change colour and lose their chlorophyll. Carbohydrate is stored in leaves rather than translocated down to the tubers, so there are fewer tubers per plant and up to 80% loss of yield. The tubers are brown inside although they look normal on the outside.

Control

The disease is controlled by good management as viruses are not killed by pesticide sprays. Only virus-free seed potatoes, called certified seed, should be planted. In the main growing regions, insecticide sprays are used when aphid numbers rise. Great care must be taken to remove all potatoes from a crop so that there are no 'strays' to act as reservoirs or overwintering quarters for the aphids.

Tobacco mosaic virus

TMV was the first virus to be isolated and have its structure established, and it is much used as a research tool. You can see its structure in Figure 2.14. However, it is also an economic problem to growers of tobacco, tomatoes and sometimes potatoes and ornamental plants.

The virus is transmitted mechanically from infected plants through abrasions and small wounds made during routine cultivation – for example, tying up tomato plants. It infects systemically, and the effects are mottled and crinkled leaves, stunting and some yellow mottling. Fruits are deformed with yellow spots and streaks.

The virus can survive in infected material for years so good hygiene is vital to control and prevent infection. It can also survive in smoking tobacco and weeds. One of the first transgenic resistances to be developed was resistance to this virus.

QUESTIONS

9.13 Suggest reasons why viruses are transmitted by vectors.

9.14 Outline three methods that could be used to control the spread of a virus infection in a potato crop.

9.15 Very briefly, explain why virus diseases make leaves lose their green colour.

9.16 TMV virus also infects tomatoes. Explain why gardeners who smoke are advised to wash their hands between smoking and handling their tomato plants. Give two other ways in which a farmer could ensure that his crop of glasshouse tomatoes remained free of TMV.

9.7 BACTERIAL DISEASES

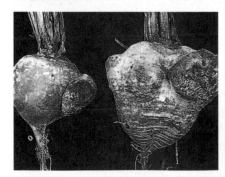

There are some very damaging bacterial diseases, but on the whole they are less important than fungi and viruses. Bacteria enter through stomata and lenticels, or through wounds. They target young dividing cells with a high nutrient content. Bacteria secrete enzymes such as pectinases to degrade the tissues, which they can then colonise. The soft rots caused by *Erwinia* species are important from the consumers' point of view because they are pests of stored potatoes and onions, but the pectinase produced by *Erwinia* is commercially useful. *Agrobacterium tumefaciens* stimulates unusual cell proliferation when it infects, resulting in galls and tumours.

FIG 9.8 *Agrobacterium* causes cell proliferation, leading to distorted growth.

9.8 HOW PLANT DISEASES ARE CONTROLLED

The best way of controlling crop infections is by reducing the opportunities for infection. A combination of different approaches is used to minimise infections and to eradicate sources of infection. Integrated pest management schemes aim to use several different approaches at the appropriate time to control the numbers of pests.

Good hygiene and weed reservoirs

All sources of infection must be removed, including remnants from previous crops. Infected material should be **burnt**. Wild host plants should be controlled so that they do not act as reservoirs of infection. Several different techniques are available. **Mulching** with black polythene or special mulches reduces weed seedlings; **weeding** and **hoeing** remove emerging weeds; and **selective herbicides** are available that kill weeds but not the crop.

Clean starting material

FIG 9.9 Callus cells are unspecialised cells which can eventually regenerate plantlets.

Only seeds, tubers, cuttings and seedlings free of viruses or fungi should be planted. Seed potatoes and plant cuttings can be supplied certified free of particular pathogens. Though fungi may pass through seeds, few viruses are transmitted through seeds or pollen, so plants grown from seed should be virus-free.

Micropropagation is the usual method of producing virus-free plants. The growing points of plants, called meristems, are regions where cells are multiplying very quickly. Viral infections may not reach these growing points or may not multiply quickly enough to infect all the new cells. The meristem cells can be separated out and those without the infection used for callus culture, which can then be used to produce large numbers of virus-free plantlets; the techniques are described in Section 9.9. Figure 9.9 shows a callus culture. Samples of the virus-free tissue can be kept at low temperatures for use at a future date.

Growing different varieties

Crop rotation, which involves growing crops on a different piece of land each year over a cycle of years, deprives soil-borne pathogens of new hosts. Spores and other resistant structures may not survive until the rotation brings susceptible plants to that area again. Monoculture encourages the evolution of pathogens – the more a particular cultivar of a crop is grown, the more likely resistance will be lost. It is better to have blocks of different cultivars grown in the same area, which limits pathogen spread and eases selective pressures. Similarly, it is better not to grow the same varieties in the autumn and spring as this again helps control the pathogen.

Vector control

FIG 9.10 Keeping aphids out by using screens with tiny mesh keeps diseases out too.

The most important virus vector in Britain and much of Europe is the peach–potato aphid. Insecticides are used to control the aphids but there are drawbacks. Many aphids and other vectors become resistant to the insecticide but other valuable insects may not, so there is an inevitable toll on ecosystems. Also, insecticide bioaccumulates in food chains. Some insecticides are more selective in their action than others but even so, spraying has to be done at times of the day when insect pollinators such as bees and hoverflies are not active.

Other techniques can be used in glasshouses used for crops or propagation. The simplest is shown in Figure 9.10. Doors, windows and ventilators are covered in mesh too small for aphids to get through.

Aphids use particular plants as winter refuges. For example, *Myzus persicae* overwinters on a range of plants including the *Prunus* family – these are the cherries and their relatives. These winter host plants may have to be removed from crop areas. There are crops that can be planted at different times of the year with staggered harvest times. As a result there are aphid food plants around in late autumn and early spring on which aphids survive to infect spring crops. Farmers may need to cooperate to prevent overlaps between different farms.

Biological control uses natural parasites or predators of the vector species to keep the numbers down to low levels. Currently there are only a few methods for controlling vectors in use but other sorts of insect pests in glasshouses are successfully controlled this way. For example, cultures of *Bacillus thuringensis*, which produces a protein toxic to caterpillars, or solutions of the toxin itself, are available to horticulturists. The gene for the toxin has been transferred into cotton and maize to control insect damage in the field. You can find out more about this toxin in Chapter 6. Heart rotting fungi that spread from tree stumps to nearby trees can be discouraged by the introduction of a harmless fungal species, which degrades only dead wood, into a tree stump or a logging wound. This prevents the more harmful species establishing itself. *Trichoderma* has been shown to be effective against a number of fungal pathogens. These biological controls are very specific, only affecting the target species and not affecting beneficial organisms. They disappear quickly and do not appear to affect animals and humans.

Chemical dusts, smokes and sprays

The only effective chemical controls are fungicides, apart from sterilising solutions to treat soils. Organo-mercury fungicides and the **dithiocarbamates**, which are organo-sulphur compounds, discovered last century can be used on a wide range of plants with few side effects.

Fungicides are used as a preventative measure, for example as seed and root dressings before planting. They are applied to growing crops once there is enough of the pathogen present and weather conditions are suitable for fungal growth. It may not be cost-effective to spray against an isolated outbreak of disease affecting only a small portion of a crop.

Systemic fungicides are absorbed by the plant and translocated to all parts of the plant, protecting wherever the fungus infects. They include compounds such as Benomyl and thiophenate. Other fungicides are **surface-active agents**, that is a dust or liquid spray lies on the surface of the plant where it kills fungal spores or hyphae growing over the surface.

Fungicides target different processes within plants. **Single-site** fungicides affect one or two metabolic activities in a pathogen whereas **multi-site** fungicides affect a wide range of activities. Most systemic fungicides are single-site, for example organo-sulphur compounds inhibit mitosis in the fungus and triazoles interfere with fungal cell wall synthesis. Multi-site fungicides often contain copper, which upsets energy production and affects many processes.

When fungicides have to be used several times in a season, it is better to use two together to minimise the development of resistance in the pathogen. They should be multi-site rather than the same single-sites all season. It is relatively easy to evolve resistance to single-site fungicides but much harder to become resistant to multi-site fungicides. All fungicides should be used with care, see Figure 9.11. There are rigorous rules for the safe handling and storage of pesticides of all types. Certain fungicides cannot be applied to edible crops close to the time of harvest because they leave residues on the plants.

FIG 9.11 Spraying fungicides and pesticides reduces crop damage but the spray could be a hazard to the workers. Workers have protective clothing, including gloves and face masks, to limit their exposure.

Growing resistant strains

Modern crop varieties have better features than their antecedents but many lack resistance mechanisms to pathogens. Resistance is usually carried genetically, for example a few genes for making a thick wax layer results in a thick cuticle that is too difficult for a pathogen to penetrate. Plants in wild populations have varying combinations of resistance genes. Selective breeding for yield has resulted in resistance and other features being lost in the genetic reshuffle that occurs whenever pollen and ovules are formed. Plant breeders have used wild species to introduce resistance genes into crop varieties. These need less pesticide treatment for more yield and so are more economic. You can read about breeding resistant strains in Section 9.10. Sooner or later though, a pathogen evolves a method of overcoming the plant resistance and new cultivars are needed.

QUESTIONS

9.17 Name an anti-fungal compound used for plant infections and explain how it works.

9.18 A farmer grows organic fruit and does not use conventional chemical sprays to control disease. Give two ways in which the farmer could reduce the chance of fungal disease in the fruit bushes.

9.19 A farmer discovered 40% of his potato crop had a virus infection that made the inside of the tubers brown and inedible. As he could not sell his potatoes he ploughed them into the ground in disgust. He thought that he could at least use them as organic material to improve the soil. He replanted with new seed potatoes the following year that he bought as certified virus-free stock. To his horror he found that when he lifted his second year's crop the potatoes were infected again. Explain three ways in which his virus-free potatoes could have been infected.

9.9 PLANT BIOTECHNOLOGY

Plant breeders and producers have two major areas of activity. One is to produce better strains of crop plants that have a better yield or are more resistant to pests or adverse weather conditions. You can read about this is Section 9.10. The other is to produce large numbers of horticultural and agricultural plants with desirable features quickly. Plant tissues are much more versatile sources of new plant material than animal tissues; even single cells can be used to regenerate an entire plant. Plant cells can be grown as protoplasts, which are plant cells without cell walls, as callus culture or as explants of particular tissues depending on their eventual purpose.

Micropropagation

Breeders may have developed a single plant, or a small group of plants, with a particular combination of desirable features. Seeds from these plants may not have the same features because of the random nature of meiosis and the contribution of the other parent plant. Taking cuttings will eventually build up a stock of genetically identical plants. This is relatively slow and some plants will not propagate by cuttings very easily, or only at certain times of the year. The biotechnological technique of **micropropagation** offers a much quicker way of generating very large numbers of genetically identical plants from a single stock plant. Important plants such as yams, the solanums, oil palms, winter pansies and expensive cut flowers, for example gerberas, lilies and orchids, are routinely propagated by micropropagation. The techniques are also useful for plants that are difficult to obtain as seed, such as bananas.

The technique exploits the ability of plant tissues to regenerate new tissues that differentiate into different types of cells. A suitable parent plant is selected and

FIG 9.12 Huge numbers of identical plants can be generated from a single parent plant carrying desirable features.

small pieces of tissue are used as a source of **explants**. Usually, dividing tissues such as buds, hypocotyl or shoot tips of a single plant or seedling are used for the explants but other tissues can be used. These are placed in a sodium hypochlorite sterilising solution to kill microorganisms on the surface, then sterile instruments are used to cut them up into smaller sections. These are placed individually into small tubes of sterile agar containing sucrose, which replaces photosynthesis, and plant growth regulators that stimulate cell division and differentiation. Cells in the explant divide under the influence of the regulators, including many cells that do not normally divide in a whole plant. A mass of unspecialised cells, or **callus**, develops within a few weeks. The callus tissue and small samples taken from it can regenerate a new plant genetically identical to the explant source. Small plantlets, like those shown in Figure 9.12, develop from the callus under the influence of the hormones. A large number of genetically identical plants, or **clone**, is built up very quickly. The little plantlets are separated and potted on or used as a source of more tissue for more explants.

This method generates hundreds of thousands of plantlets a week without the need for huge areas of land. Another advantage is that it allows far more infection-free plants to be raised per unit area of bench space than ordinary methods. The horticultural companies may have to provide expensive equipment, pay attention to sterility and develop special staff skills but conventional propagation is also demanding of equipment and hygiene.

9.10 BREEDING RESISTANCE INTO PLANTS

A plant's resistance to disease is usually genetic; a pathogen's ability to infect a plant is also due to the genes it carries. A wild plant population is a mixture of genetically different individuals with different resistance genes. Pathogens within a population are also genetically varied and have different 'infection' genes. Plants with good resistance mechanisms survive; pathogens with the best mechanisms for overcoming resistance also survive. The plants and pathogens are in an evolutionary struggle where each evolves defensive and invasive mechanisms in response to the other.

Ideally a crop strain has more than one resistance gene, which makes it more difficult for the pathogen to overcome the resistance. Knowing about the infection process gives us clues to how to obtain more resistant strains of plants.

Several resistant strains should be available so that local farmers can grow different strains to minimise pathogen evolution. However, all farmers want the best possible yield, so the strains would also need to have similar growth characteristics, so that no-one is disadvantaged by growing one strain rather than another.

Selective breeding

In traditional selective breeding, strains are crossed using pollen transfer between a plant with the desired disease resistance to another with good growth, and collecting the seed produced. The seeds are grown and the performance of the individual plants assessed. Good possibilities are bred on or crossed again until, several years and millions of plants later, a strain is produced which has the desired disease resistance and good growth.

The sources of disease resistance genes are:
■ varieties of the crop grown in other parts of the world,
■ wild relatives and antecedent strains,

- artificial induction by treating seeds with a mutagen such as irradiation or particular chemicals. This generates a number of variant plants in a short time to use in selective breeding.

Micropropagation offers a shortcut to crop improvement. Cells in a callus culture can each develop into a plant. However, when individual cells are used to regenerate plants, some of the plants produced do have variations in their features even though they are genetically identical. These plants can be checked and selected or discarded quickly. Far fewer plants need to be grown and allowed to set seed than in selective breeding. It also conserves favourable gene clusters. Once a desirable strain has been produced and tested, it can be propagated very quickly, shortening the time taken to get from trials in the research institute to the farmer.

Protoplasts, pollen and hybridisation

Cell culture techniques can produce variant plants for crop improvement. Protoplast culture is a technique of growing plant cells in culture without their cell walls. To make protoplasts, plant cells have their cell walls removed by gentle enzyme action, usually a mix of cellulases and pectinases. The cells are cultured in a well-regulated medium to avoid osmotic shock. They will quickly secrete new cell walls and regenerate plantlets. The surprising thing is that the young plants are often quite different to each other, even though they are all from the same parent. This type of variation is called **somatic variation**. The technique produces a large number of variant plants and is useful for species such as lettuce, which are difficult to pollinate artificially in crossbreeding. It is also quick and conserves favourable gene clusters.

Pollen cultures can be used to generate lines of plants that are homozygous for every character; these are useful parent lines. The male gametic nuclei inside pollen grains are haploid – they have a single set of chromosomes. When these are treated with colchicine, mitosis is affected and the duplicated chromosomes stay together as a pair, giving cells which are homozygous for every gene. They can be cultured and used to regenerate homozygous plants for use in breeding.

Genetic modification

The techniques outlined in Chapter 6 dramatically reduce the time taken to introduce resistance genes because the desirable genes can be transferred directly into plant cells. These techniques are particularly useful where target plants have a long life cycle, or do not reproduce well sexually, or genes are wanted from unrelated species that would not crossbreed in normal circumstances.

Plants have been modified in a number of ways to improve crop productivity:

- by reducing crop losses through pests, diseases and competition from weeds,
- by improving shelf life or storage of a product,
- by improving the weather resistance of a crop plant.

The first genetically modified crops include

- maize that has been genetically modified to carry a gene whose product is toxic to insect pests. The insects are killed as they try to feed on the maize stems;
- soy beans and cotton have been genetically modified for resistance to herbicides. Crop plants that can resist herbicide would be useful because farmers could spray actively growing crops to remove the competing weeds.

Genes have been transferred into broad-leafed plant cells using the plasmid carried by *A. tumefaciens*. The technique works well with tomatoes and potatoes. Transfer of insect toxin into maize, which belongs to the cereals and grasses, used the technique of bombarding plant cells with DNA-coated pellets. The resistance genes have been derived from a virus, a bacterium and a petunia.

A tomato cultivar uses **anti-sense technology** to prevent the production of softening enzymes. A new gene makes a product that combines with the mRNA for the softening enzyme and blocks its production. The tomato now takes longer to soften but otherwise ripens as usual. The tomatoes can be picked later in ripening yet still get to the shops or factory in good condition. There is research into using crops to make novel useful raw materials, such as polyhydroxyalkanoic acids used to make biodegradable plastics (see Chapter 5).

There are concerns that genes for herbicide resistance could spread via pollen into weed plants, making them more difficult to kill, and also that farmers may be tempted to spray more than usual, killing a wider range of wild plants. Other concerns centre on whether these new gene products could harm humans consuming them, or enter food chains through animals eating them. All such products have to undergo a series of tests to demonstrate safety, though again there is concern that the tests might not be rigorous enough, or that damage may occur during the testing process.

QUESTIONS

9.20   What are the advantages of using cloned plants instead of collecting seeds?

9.21   Construct a flow chart to show the process of micropropagation. Reread the passage on growing plant cells then outline the procedure you would use to produce a clone of violet plants for a mini-enterprise scheme in time for Mother's Day in March.

9.22   You have been provided with a diseased plant and several healthy plants of the same variety. You also have a pure culture of a fungus that you suspect is causing the disease. What investigations could you carry out to show whether or not the fungus you have is the cause of the plant disease?

Exam questions

9.23   *Bacillus thuringiensis* produces a protein that is toxic to leaf-eating caterpillars. This protein has been used by farmers as natural insecticide. Recently, the gene that codes for the toxin has been genetically engineered into several crop plants.
   (a)   Outline how the gene that codes for the toxin could have been isolated. (5)

The gene, once isolated, is inserted into a host plant cell either using the bacterium *Agrobacterium tumefaciens* as a vector to infect a plant cell or using a particle gun which shoots DNA-coated pellets into a plant cell.
   (b) (i)   Suggest why a plasmid vector cannot be used to insert the gene into plant cells. (1)
       (ii)   Explain why it is important to insert the gene into a single isolated plant cell rather than into a cell within a whole plant. (2)
   (c)   State two environmental implications of genetically engineered pest resistance in plants. (2)

It has been suggested that the introduction of the gene for toxin production from *B. thuringiensis* into a wide range of crop plants could result in a loss of the effectiveness of the toxin.
   (d)   Outline how this loss of effectiveness of the toxin might occur. (2)

OCR Sciences, March 1999, Paper 4806, Q. 3

9.24   Some wild varieties of potato have a natural resistance to potato blight, a disease caused by the parasitic fungus *Phytophthera infestans*.
   (a)   Explain how this resistance may have evolved. (4)

The potato plant stores food materials in swollen regions of its underground stem, known as tubers. Spores of the fungus land on the leaves and germinate to produce threadlike structures known as hyphae. These penetrate the tissues of the leaves and, eventually, the tubers, releasing enzymes which allow them to penetrate the cells. Once established inside a cell, the hyphae release further enzymes which kill the cell and the fungus can then absorb the digested contents.
   (b)   Suggest **two** ways in which the hyphae might penetrate healthy leaves. (2)

(c) State the type of enzyme produced by the hyphae to penetrate (i) the cell wall, (ii) the cell membrane. (2)

(d) Explain how infection by the fungus will affect the crop yield of the plant. (3)

Researchers have been able to use a bacterium to insert a resistance gene, known as a 'suicide' gene, into potato cells.

(e) Describe how the 'suicide' gene is introduced initially into the bacterium. (4)

OCR Sciences, June 1999, Paper 4802, Q. 2

9.25 (a) Describe, using one named example from agriculture, how an organism may be genetically engineered. (9)

(b) Discuss, using specific examples, whether the potential benefits of genetic engineering outweigh the potential hazards. (7)

OCR Sciences, June 1999, Paper 4807, Q. B2

SUMMARY

Fungi and viruses are the most important plant pathogens.

Pathogens cause tissue damage, kill cells, produce toxins or affect plant activity, reducing the ability to photosynthesise, translocate sugars, transpire or store nutrients.

Pathogens are transmitted by wind, water, soil, or inoculated directly into sap. Vectors transmit viruses. Infections spread from continent to continent and reduce the yield of major crops.

Plants have natural barriers to pathogen entry and can induce chemicals to inhibit pathogen activity.

Control measures aim to decrease the transfer of pathogens. Fungicides are used to treat fungal infections. Multi-site fungicides affect a range of metabolic processes; single-site fungicides only affect one or two.

Viruses are controlled by hygiene and vector control. Reservoirs of infection in wild and infected crop plant material must be removed. Insect vectors are controlled by insecticides, removal of refuges and by avoiding times when insects are prolific.

Resistant strains of plants are being bred.

Micropropagation is used to generate huge numbers of identical plantlets.

Genetic modification is used to transfer resistance genes into plants.

Examination Questions

1 Laboratory experiments were carried out to investigate two methods of producing an antibiotic by the bacterium *Streptomyces lavidans*. In the first method, the antibiotic was produced in a series of batch fermentations. Each fermentation took 6 days. Two days were needed to clean and sterilise the equipment before the next batch was started. Nine batches were grown over 72 days. In the second method, a continuous fermentation process was run for 72 days. The results of the experiments are shown in the table.

FERMENTER TYPE	DRY MASS OF BACTERIA PER LITRE OF FERMENTATION LIQUID IN GRAMS	TOTAL ANTIBIOTIC PRODUCTION OVER 72 DAYS IN ARBITRARY UNITS	SPECIFIC PRODUCTIVITY: ANTIBIOTIC PRODUCTION IN UNITS PER GRAM OF BACTERIA PER DAY
Batch	5.6	29.25	0.073
Continuous	3.2	21.88	

(a) (i) Calculate the specific productivity for continuous production. (1)

(ii) As a result of these experiments, it was decided to produce this antibiotic by continuous fermentation. Using the data in the table, explain why. (2)

(b) If this antibiotic were produced commercially, the cost of production would be important, as well as the rate of production. Suggest why the cost of production for batch fermentation might be higher than for continuous fermentation. (2)

(c) The removal of material at intervals during the continuous fermentation process could lead to a loss of bacteria. Explain how immobilisation may prevent this loss. (2)

AQA (NEAB) Biology, March 1999, Paper BY06, Q. 5

2 The table shows some of the events in the history of vaccine development.

DATE	VACCINE PRODUCED
1796	smallpox
1885	rabies
1900s	typhoid, cholera
1914	tetanus
1920s	tuberculosis
1930s	diphtheria, yellow fever
1940s	influenza, whooping cough
1955–60	poliomyelitis
1960s	measles, rubella
1968	mumps
1970s	chickenpox
1980s	hepatitis B, combination vaccine (measles, mumps and rubella)

(a) (i) Give **three** vaccines from the table which are normally given to children in the UK in the first year of life. (1)

(ii) Give **two** vaccines from the table which have been developed in the last 50 years and which are **not** included in the schedule of immunisation of children in the UK. (1)

(b) (i) Explain how a vaccine prevents the development of a disease. (3)

(ii) Explain why some of the vaccines in the table are more effective in preventing the development of disease than others. (2)

AQA (NEAB) Biology, June 1998, Paper BY08, Q. 5

Appendix A: Ideas for practical and other investigations

Safety

If you are planning a practical investigation, it is vitally important that you discuss your procedure, choice of organisms and growing temperatures with your tutor to assess risk and avoid potentially hazardous activities. It is generally unsafe to consume any item made in a laboratory, for example yoghurt. Some antibiotics produce allergic reactions; these should be used as impregnated discs. You should not open cultures you have grown; dispose of them safely according to your laboratory procedures. Check with your tutor that a suitable safe organism is available to you.

Note: (ICT) indicates that you could use datalogging equipment to monitor changes.

Basic techniques

You should be able to:

1   Pour a nutrient agar plate.

2   Streak a sample of a pure culture of bacteria onto a nutrient agar plate.

3   Inoculate a nutrient broth.

4   Inoculate a sample of pure culture of fungi onto a malt agar plate or other suitable medium.

5   Monitor the growth of a pure culture of bacteria in a nutrient broth using a colorimeter or light sensor to measure the increase in turbidity. (ICT)

6   Carry out a Gram stain on a bacterial culture.

Investigating lactic acid producing bacteria

1   Effect of varying salt concentration on sauerkraut production .

2   Monitor acidity changes in sauerkraut production (ICT).

3   Change in pH or viscosity during the fermentation of milk by *Lactobacillus bulgaricans* to make yoghurt (ICT).

4   Compare skimmed, semi-skimmed, sterilised and powdered milk setting qualities in yoghurt making.

5   Investigate microbial activity in milk using resazurin or methylene blue, both decolourised by microbial activity.

Yeast

1   Effects on yeast growth rate of differing temperatures or pH, monitor numbers using haemocytometers, or increase in volume of dough due to carbon dioxide production.

2   Ability of yeast to ferment a range of sugars and starches.

3   Effect of temperature on yeast fermentation of sugar solutions.

4   Investigate the best conditions for bacterial amylases to degrade grain starch.

5   Immobilise yeast in alginate, pack into columns such as a gas burette or syringe, investigate its enzyme activity by pouring sucrose solution through the column, monitor appearance of glucose using glucose detection strips.

Biodegradation and biotechnology

1   Ability of bacterial enzymes in washing powder to degrade specific materials, effect of incubation temperature, pH, enzyme concentration. (ICT)

2 Digestion of cellulose in filter paper by *Cellulomonas*.

3 Investigate bacterial proteases, e.g. by degrading coagulated egg white, or diameter of clearing around wells cut in skimmed milk agar, or filter paper discs soaked in enzyme.

4 Use of pectinase to clarify freshly pressed fruit juice (ICT).

5 Immobilise lactase in alginate beads, pack the beads in a column such as a syringe or gas burette, pour milk through and monitor conversion of lactose to subunits using glucose detection strips.

6 Micropropagation of plants.

7 Production of plantlets from callus culture and factors affecting growth rate.

8 Transfer of genes using *Agrobacterium*.

The National Centre for Biotechnology Education at the University of Reading has developed a number of practical activities and packages for schools, including gene transfer activities. Their website lists some activities, and worksheets can be downloaded.

More information

■ National Centre for Biotechnology Education: www.ncbe.reading.ac.uk

Fungi

1 Investigate fungal growth rates by measuring mycelial diameter.

2 Investigate antibiotic activity using discs on a lawn of bacteria.

3 Make soy sauce or bean curd.

4 Survey fungal types in local environs.

5 Investigate spore formation in various fungal species (note: asthmatics – special care).

6 Growth of fungi used in blue cheese making.

7 Cellulase activity by fungi.

8 Degradation of starch in starch agar by fungi.

Health and disease

1 Effectiveness of fungicide against damping-off of seedlings in a greenhouse or cold frame.

2 Sensitivity of monoclonal antibodies in pregnancy testing kits, if HCG is available.

3 Progress of growth of *Monilinia* on apples in warm conditions.

4 Bioassay of the effectiveness of disinfectants and antiseptics. Measure how bacterial growth is inhibited by comparing the diameter of the zone of clearing around filter paper discs soaked in, for example, anti-bacterial spray cleaner, toothpaste, deodorant, on a lawn of bacteria in nutrient agar.

5 Lowest concentration of disinfectant required to kill a bacterium growing in nutrient broth, monitoring by turbidity. If the organisms are killed, there will be little change in turbidity. (ICT)

Other

1 Investige the growth requirements of particular bacteria.

2 Look at the effect of length of exposure to UV light on bacteria spread on a nutrient agar plate – caution: safety hazard using UV.

3 Examine the effect of heat shock on bacterial viability. Use a small volume of bacterial culture and hold at temperatures above 55 °C for 15 min. After cooling, add nutrient broth and look for time for turbidity to develop.

4 Investigate the range of protozoa or unicellular algae in a hay infusion (dried grass cuttings kept in tap water for 3 days).

5 Growth of pond photoautotrophs with light of differing wavelengths.

6 Pigmentation in cyanobacteria with light of differing wavelengths.

Marketing and information

1 Investigate the range of microbial products available in the local supermarket.

2 Survey the acceptibility of mycoprotein to consumers. Organise a taste test using mycoprotein pate or stir-fried Quorn..

3 Survey immunisation rates among fellow students and investigate attitudes towards immunisation.

4 Plan a campaign to improve the uptake of vaccination against meningitis C amongst year 13 students planning to enter higher education. Produce an artwork brief for campaign materials. If you study psychology as well, you could consider what factors are most likely to encourage people to follow health advice.

5 Monitor newspaper reports of food poisoning outbreaks and the assigned causes. Produce an information leaflet on how to prevent food poisoning at home.

6 Design a leaflet to reduce the transmission of a common disease such as hepatitis or salmonella.

7 Survey the use of anti-microbial compounds in food preservatives.

Appendix B: Glossary of biological terms

This glossary covers terms used in the book that are familiar to those who have studied A-level biology but may be unfamiliar to new students or those on other courses. Many other terms are defined within the text when they are first used, and marked by **bold type**.

A

Adhere/adhesion — the mechanisms by which bacteria make links between each other and the surface they are growing on so they are not washed away by fluid currents or impacts from moving particles.

Adjuvant — something that boosts an immune reaction. Aluminium hydroxide absorbs antigen particles, but other adjuvants act in different ways.

Adsorb — small powdery particles may be able to take up, or adsorb, substances on their surfaces.

Algal bloom — algae (and also cyanobacteria) reproduce at a rapid rate when given very favourable conditions for growth, faster than they are eaten. The numbers are so great that they block light transmission through water. It is usually due to excess nitrates and phosphates draining into warm waters, from fertiliser or water treatment plants.

Ambient temperature — the temperature of the surrounding air / water.

Assay — a biological assay is a way of assessing how much of an active substance is present, but using an indirect measure such as an effect on something else. For example, the relative strength of antibiotic solutions can be compared by measuring the zone of inhibition around antibiotic soaked discs placed on a film of bacteria growing on a nutrient agar plate.

ATP — adenosine triphosphate. It is made during respiration when glucose is broken down. It is a store of chemical energy within cells and used as the immediate source of energy for chemical reactions.

B

Bioaccumulation — a process in which a potentially harmful substance is taken into animals or plants at the base of a food chain and not broken down or excreted. These are eaten by animals further up the food chain, which accumulate larger quantities as they continue to take in their food items. Predatory animals higher in the food chain accumulate harmful quantities.

C

5-Carbon sugar — a small ring-shaped carbohydrate molecule, like glucose but with five carbon atoms in the molecule not six.

Clinker — fused ash left in a furnace after coal or coke is burnt.

Coagulation — occurs when proteins or protein-rich materials such as cytoplasm change physically from a relatively fluid state to a solid. For example, the change that occurs when an egg is boiled. It can be brought about by heat, changes of pH, or certain chemicals.

D

Debilitate — the body systems become weakened, often due to continuous ill-health or malnutrition.

Differentiation — the changes in structure and activity a cell undergoes between its production by cell division as an unspecialised cell and its final adaptations for a particular function.

E

Electrolytes — ions dissolved in body fluids, mainly sodium, potassium and hydrogen carbonate.

Emulsion — a suspension of one substance in another, into which it does not usually dissolve. Examples include oil and water in salad dressings, and proteins in water in cytoplasm.

Endocytosis — a process in which materials are taken into cells. The outer cell membrane folds inwards and a small bubble forms, containing whatever is being taken in. The bubble is pinched off and passes into the cell interior.

Exponential growth — the stage at which microorganisms are reproducing so rapidly that the numbers are increasing logarithmically.

G

Genetic probe — a stretch of single-stranded DNA or mRNA which can bind to a corresponding sequence in a sample of DNA. The probes carry a marker such as radioactive isotope atoms, so the exact position where they bind to the DNA sample can be located. Used to locate the position of genes or specific DNA sequences.

Gram-positive/negative — the response of bacteria to a staining technique developed by Gram. Used to classify some bacterial groups as it reflects cell wall structure (see Chapter 2 for details).

H

Hydrolyse — to break down a molecule into two smaller molecules by interaction with water and an enzyme or heat or an acid.

I

Indigenous population — the people or animals normally found in a particular area, not including colonists or their descendents.

L

Lysis — the splitting or breaking open of cells.

M

Maternal antibodies — antibodies produced by pregnant women which pass to the baby either across the placenta or via breast milk.

Metabolism — the physical and chemical processes going on within cells making and degrading materials. A substance produced by these activities is a metabolite.

Motile — used of microorganisms that can propel themselves around in fluids.

Mucous membranes — the tissues that line cavities and tubes in the body such as nasal passages, etc. Also referred to as mucosa.

Mutualism — a relationship between two species of organism in which both benefit from the association.

N

Notifiable disease — doctors have to record any cases of the disease coming to their attention in a central database.

P

Polysaccharide — a large polymer molecule made of repeating sugar subunits, e.g. starch or cellulose.

Precursor — most substances made within a cell are the result of a linked chain of chemical reactions starting with raw materials. Precursor molecules are molecules produced or used in one of the intermediate steps to make the final product.

Productivity — the rate of production of new material, generally measured in mass increase per time period.

Prokaryote — an organism with prokaryotic cell features, mainly bacteria and cyanobacteria, see Chapter 1.

R

Repressed genes — genes which are present on DNA but not functioning within a cell. A mechanism within the cell, such as a repressor molecule attached to the start of the gene, blocks the action of enzymes needed to start the process of making the gene product.

Resolving power — a measure of the smallest object that can be distinguished by a microscope.

Restriction enzyme — an enzyme that can cut DNA molecules into smaller fragments. Each restriction enzyme cuts within a particular base sequence, leaving fragments with known base sequences at each end.

S

Secrete — the process in which cells synthesise a desired substance such as an enzyme or hormone and pass it out of the cell. Waste substances made during metabolism are excreted from cells.

Selection pressure — factor in the environment, such as a predator or climatic factors, that affects the survival of individuals displaying certain features.

Stock culture — a standard culture of microorganisms, kept in storage and used as a source when a new culture of microorganisms is needed for a process. It ensures consistency.

Substrate — two definitions: (a) the surface on which a microorganism lives; (b) the substance used in a chemical reaction catalysed by enzymes.

T

Trace elements — minerals required in very small quantities (milligrams or less) by living organisms, e.g. cobalt, manganese, zinc.

Tumour — a cluster of unspecialised cells developing into a mass of unspecialised tissue without a proper function in the tissue where it is growing.

W

Water potential — a measure of the concentration of water in a cell.

Extending your understanding

You can widen your knowledge and deepen your understanding by visiting a reference library (don't overlook your own school or college library).

Many of the books written for post A-level students are easy to understand and provide a bit more depth on individual topics. There are many books on biotechnology written for the interested public. Many are classed as microbiology books but you will also find good texts in sections on plant pathology, drugs and therapeutics, public and environmental health, food technology, agriculture, biomass energy and waste management.

Reference libraries usually carry specialist magazines. Almost every issue of *New Scientist* and *Scientific American* carries articles relevant to your study areas, so does *Biologist*, the journal of the Institute of Biology. For those interested in health studies, *World Health*, the magazine of the World Health Organisation, carries many interesting articles, written for the ordinary reader. If you live far from a reference library, don't despair. The Internet gives you access to many magazines' websites where you can read articles from previous issues and sometimes download them.

Wider reading

If you are interested in any of the topics in this book, the books listed below will tell you more:

The Biochemical Society publications:
> *Enzymes and their Role in Biotechnology*
> *Immunology*
> *Recombinant DNA Technology*

Biotechnology and Biological Sciences Research Council publications
The Biotech Century: Harnessing the Gene and Remaking the World. Jeremy Rifkind.
The Coming Biotech Age. Richard W. Oliver.
Essentials of Food Microbiology. John Garbutt, Arnold Publishing.
How the Immune System Works. Lauren Sompayrac, Blackwell.
Invisible Allies. Bernard Dixon, Temple Smith.
Jenner's Smallpox Vaccine. Baxby, Heinneman.
Life at Small Scale. D. Dusenbery, Scientific American Library.
Man Made Life. Jeremy Cherfas, Blackwell.
Microbes and Man. John Postgate, Cambridge University Press.
Of Mice, Men and Microbes. Harper & Meyer, Academic Press Inc.
Milestones in Microbiology. T. D. Brock, ASM Press.
Pasteur – The History of a Mind. Duclaux, Library of New York Academy of
> Medicine/Scarecrow Reprints.
Pirates of the Cell. Scott, Blackwell.
Plagues and Peoples. W. H. McNeill, Blackwell.

Web searches

There are a host of useful websites concerning aspects of microbiology and biotechnology. Make your search tightly targeted at the topic you want to know about. For example, use the name of a bacterium used to extract copper, not just copper; or biogas not compost; or waste water not just sewage. A tip to help you locate a reliable website on (say) vinegar production rather than someone's opinions of vinegar on chips is to look at the web address. British, American and Australian Universities end in edu. (USA), ac.uk (UK), edu.au (Australia). They have departmental websites with factual information about topics within this book. Official bodies such as the Environment Agency usually have web addresses ending in org.uk. Government websites end in gov.uk. Companies involved in making or selling microbiological products also have useful websites.

The following have useful sites.

Babraham Institute (gene work): www.bi.bbsrc.ac.uk

Bioscience Network: www.bioscinet.bbcsr.ac.uk

Biotechnology Knowledge Centre in Iowa: www.biotechknowledge.com

Cells Alive: www.cellsalive.net

The Centre for Alternative Technology: www.cat.org.uk

You might enjoy a trip around the microbe zoo at the **Digital Learning Centre for Microbial Ecology**: www.commtechlab.msu.edu/sites/dlc-me/zoo/

The Environment Agency – check the water quality near your home: www.environment-agency.gov.uk

Food and Agriculture Organisation: www.fao.org

Institute of Food Science and Technology: www.ifst.org

National Centre for Biotechnology Education: www.ncbe.reading.ac.uk

The National Health Museum, USA, genetic engineering: www.accessexcellence.org

Public Health Laboratory Service: www.phls.co.uk

Roslin Institute – cloning and other genome work: www.ri.bbsrc.ac.uk

Sanger Institute – human genome project: www.sanger.ac.uk

Science Daily is a digest of latest discoveries: www.sciencedaily.com

Science Traveller International provides links to other waste water websites through www.scitrav.com/wwater/

The Society for General Microbiology has news items and useful links at www.socgenmicrobiol.org.uk

The University of Bath: www.bath-sci.co.uk

The University of East London and the University of Florida (web tours of water treatment works): www.uel.ac.uk and www.ufl.edu

US Department of Energy Biofuels Programme: www.biofuels.nrel.gov

US Food and Drugs Agency, Bad Bug Book: http://vm.cfsan.fda.gov/~mow/

The US National Dairy Council: www.nationaldairycouncil.org

World Health Organisation: www.who.int

A site which will direct you to lots of interesting sites: www.biotechsupportindia.com

Some good pictures in the **University of Nijmegen gallery**: www.microbiol.sci.kun.nl/microbiol/index.html

Magazines

New Scientist: www.newscientist.com

Scientific American: www.sciam.com

Index

Index